# The International Arms Trade

# Issues in World Politics Series

*James N. Rosenau and William C. Potter, consulting editors*

*Other Publications in the Series:*

**Journeys through World Politics**
*Autobiographical Reflections of Thirty-four Academic Travelers* (1989)
Joseph Kruzel and James N. Rosenau

**Global Changes and Theoretical Challenges**
*Approaches to World Politics for the 1990s* (1989)
Ernst-Otto Czempiel and James N. Rosenau, editors

**International/Intertextual Relations**
*Postmodern Readings of World Politics* (1989)
James DerDerian and Michael J. Shapiro, editors

**The International Politics of Television** (1990)
George H. Quester

**The New Regionalism in Asia and the Pacific** (1991)
Norman D. Palmer

**The Place of Morality in Foreign Policy** (1991)
Felix E. Oppenheim

**The Scientific Study of Peace and War** (1992)
John A. Vasquez and Marie T. Henehan, editors

**The Global Philosophers:**
*World Politics in Western Thought* (1992)
Mark V. Kauppi and Paul R. Viotti

# The International Arms Trade

Edward J. Laurance

LEXINGTON BOOKS
*An Imprint of Macmillan, Inc.*
New York

Maxwell Macmillan Canada
Toronto

Maxwell Macmillan International
New York   Oxford   Singapore   Sydney

*Library of Congress Cataloging-in-Publication Data*

Laurance, Edward J.
  The international arms trade / Edward J. Laurance.
    p.   cm.
  Includes index.
  ISBN 0-669-19928-1
  1. Defense industries.   2. Arms transfers.   3. Weapons industry.
4. Military assistance.   I. Title.
HD9743.A2L34   1992                                    92-9727
382'.456234—dc20                                          CIP

Copyright © 1992 by Edward J. Laurance.

All rights reserved. No part of this book may be reproduced or transmitted in any form or by any means, electronic or mechanical, including photocopying, recording, or by any information storage and retrieval system, without permission in writing from the Publisher.

Lexington Books
An Imprint of Macmillan, Inc.
866 Third Avenue, New York, N.Y. 10022

Maxwell Macmillan Canada, Inc.
1200 Eglinton Avenue East
Suite 200
Don Mills, Ontario M3C 3N1

Macmillan, Inc. is part of the Maxwell Communication Group of companies.

Printed in the United States of America

printing number
1  2  3  4  5  6  7  8  9  10

# Contents

List of Tables   vii

List of Figures   xi

Preface   xiii

1. **An Introduction to the Study of International Arms Transfers as a Concept and Issue Area   1**

    Arms Transfers as an Important Issue Area   1
    The Evolution and Sociology of Knowledge of the Study of International Arms Transfers   5
    Questions, Propositions, and Puzzles   12

2. **The Conceptualization and Measurement of International Arms Transfers   16**

    The Arms Transfer as the Basic Unit of Analysis   16
    Generating Reliable and Valid Arms Transfer Data: Does It Make a Difference?   20
    The Data   23
    Measurement Problems   30
    Generating Cases and the Problems Associated with Aggregating Currently Available Data   40
    Methodological Challenges   41

3. **International Arms Trade as Systems and Regimes   45**

    The Utility of the Systemic Approach   45
    Choosing among Competing Approaches and Frameworks   47
    Analyzing International Arms Trade Systems Using the Holsti Framework   48

4. **The Interwar Arms Trade System   58**

   Boundaries and Environment   58
   Characteristics of Units   62
   Structure and Stratification   64
   Modes of Interaction   69
   Regime Norms and Rules   70

5. **The Evolution of Arms Trade Systems in the Postwar Period   77**

   The Postwar Bipolar System, 1946–66   77
   The Expanding Arms Trade System, 1966–80   99

6. **The Declining Bipolar Arms Trade System, 1980–92   124**

   The Systemic Contradictions of 1980–85   124
   Boundaries and Environment   125
   Characteristics of Units   128
   Structure and Stratification   133
   Modes of Interaction   147
   Regime Norms and Rules   155

7. **Explaining National Arms Transfer Behavior and System Transformation at the Systemic Level   170**

   Explaining Supplier and Recipient Behavior at the Systemic Level   170
   Explaining Changes in Behavior across Temporal Systems   172
   Continuity, Change, and System Transformation   190
   Sources of System Continuity and Transformation and the Post-1992 Arms Trade System   196
   The Next Steps in Explaining the Arms Trade   205

**Notes   207**

**Bibliography   227**

**Index   239**

**About the Author   247**

# List of Tables

**Table 2-1.** Sample of Multiattribute Utility Scores  39

**Table 4-1.** Summary of Interwar Dependence Levels by Weapons System  63

**Table 4-2.** Combat Aircraft Exported, 1930-39  64

**Table 4-3.** Market Shares for Tanks, Warships, and Combat Aircraft, 1930  65

**Table 4-4.** Export of Arms and Ammunition of Principal Exporters, 1929-38  67

**Table 4-5.** Interwar Supplier-Recipient Patterns by Acquisition Style  68

**Table 4-6.** Modernity Indices—Aggregate Comparison of Interwar and Postwar Systems  69

**Table 4-7.** Transfers Accounted for by Licensing, Interwar Suppliers  70

**Table 4-8.** Summary of Interwar Arms Trade System  75

**Table 5-1.** Summary of Postwar Dependence Levels by Weapons System  83

**Table 5-2.** Arms Transfers and GNP of Major Industrial States, 1960-64  84

**Table 5-3.** Leading Arms-Importing Countries, 1961-65  86

**Table 5-4.** Postwar Market Shares for Tanks, Warships, and Combat Aircraft  87

**Table 5-5.** Market Shares of Leading Exporting Countries, 1961-71  88

### List of Tables

Table 5-6. Shifts in Supplier Market Shares, 1950-54 to 1960-64  88

Table 5-7. Examples of Types of Weapons Systems Traded in 1964  91

Table 5-8. U.S. Mode of Payment—FMS, Commercial, and MAP, 1963-66  93

Table 5-9. Summary of 1946-66 Arms Trade System  98

Table 5-10. Distribution of Global Arms Imports, 1963-67 to 1976-80  102

Table 5-11. Third World Weapons System Production, 1966 versus 1980  105

Table 5-12. Number of Third World Countries Importing Major Weapons from More than One Supplier Grouping, 1971-85  106

Table 5-13. Top Twelve Arms Recipients, 1963-67 to 1976-80  108

Table 5-14. Examples of Types of Weapons Systems Traded in 1976  109

Table 5-15. Soviet Hard Currency Sales, 1970-81  111

Table 5-16. Soviet Arms Transfers to Major Clients, 1975-79  111

Table 5-17. Examples of Bloc Changes in Arms Supply, 1968-77  116

Table 5-18. Summary of 1966-80 Arms Trade System  121

Table 6-1. Ranking the Second-Tier Suppliers, 1980-87  131

Table 6-2. Exports by Selected Suppliers to Countries at War as as Percentage of Total Exports of Major Conventional Weapons, 1980-98  132

Table 6-3. Number of Suppliers in the System, 1975-88  135

Table 6-4. Number of Recipients in the System, 1975-90  135

Table 6-5. Sole- and Predominant-Supplier Relationships, 1976-88  139

Table 6-6. Aircraft and SAMs Supplied to the Developing World  143

Table 6-7. Proliferation of SAMs and Air Defense Aircraft, 1980-89  143

Table 6-8. Proliferation in Production of Naval Systems, 1979-88   146

Table 6-9. European Missile Trends, 1946-87   149

Table 6-10. Nunn Amendment Cooperative Research and Development Funding, Fiscal Years 1986-92   151

Table 6-11. A Typology of Methods Employed in the Illegal Arms Trade   154

Table 6-12. Summary of 1980-92 Arms Trade System   167

# List of Figures

**Figure 2-1.** Excerpt from *World Military Expenditures and Arms Transfers 1989*   28

**Figure 2-2.** Excerpt from *SIPRI Yearbook 1990*   29

**Figure 2-3.** Excerpt from DSAA Publication *Foreign Military Sales, Foreign Military Construction Sales and Military Assistance Facts,* September 1989   31

**Figure 2-4.** Excerpt from CRS Publication *Trends in Conventional Arms Transfers to the Third World by Major Suppliers, 1982-1989,* August 1990   32

**Figure 2-5.** Sample Entry from Arms Trade Register, *SIPRI Yearbook 1990*   33

**Figure 4-1.** Interwar Arms Trade Relationships   60

**Figure 5-1.** Arms Exports to the Third World, 1951-85   78

**Figure 5-2.** Arms Deliveries to the Third World, 1963-86   79

**Figure 5-3.** Comparison of Postwar and Interwar Arms Trade Relationships   80

**Figure 5-4.** Arms Imports Ratios of Developed versus LDC Recipients, 1961-71   85

**Figure 5-5.** Postwar Supplier-Recipient Patterns, to 1968   89

**Figure 5-6.** Sole and Predominant Supplier-Recipient Relationships, 1951-85   89

**Figure 5-7.** Regional Distribution of Deliveries of Major Weapons Systems to the Developing World, 1951-65   90

xii • List of Figures

**Figure 5-8.** MAP Grants as a Percentage of Total U.S. Arms Deliveries   92

**Figure 5-9.** Third World Major Weapon Production as a Share of Total Major Weapon Imports, 1952-82   101

**Figure 5-10.** Number of Postwar Arms Suppliers   103

**Figure 5-11.** Market Shares of Global Arms Deliveries, 1965-84   104

**Figure 5-12.** Number of LDCs Receiving Major Weapons Systems, 1951-85   107

**Figure 5-13.** Regional Distribution of Arms Deliveries, 1963-87   107

**Figure 5-14.** Growth of U.S. Commercial and Credit Exports   110

**Figure 5-15.** Value of Production of Major Weapons in the Third World, 1950-84   112

**Figure 5-16.** Selected Multinational Weapons Projects and Export Sales, 1965-83   117

**Figure 6-1.** Arms Exports (Deliveries) as a Percentage of Total Exports, 1978-88   129

**Figure 6-2.** Arms Deliveries and Agreements in the 1980s   134

**Figure 6-3.** Arms Deliveries and Agreements to the Developing World in the 1980s   136

**Figure 6-4.** Market Share of Second-Tier Suppliers, 1977-90   137

**Figure 6-5.** Market Shares of Arms Deliveries to the Iran-Iraq War   138

**Figure 6-6.** Regional Distribution of Major Weapons to LDCs and Global Arms Deliveries in the 1980s   140

**Figure 6-7.** Number of Countries in the Third World with Selected Weapons Systems, 1950-85   141

**Figure 6-8.** Trends in U.S. Arms Transfers (Commercial versus FMS), 1980-89   148

**Figure 6-9.** Trends in U.S. Commercial and Credit Arms Sales, 1980-89   148

**Figure 7-1.** Offsets as a Percentage of Contract Value, 1980-87   176

# Preface

One of the first things I did as I sat down to write this book was to pull off of my shelf the fifteen major books that had been written since 1969 on the subject of the international arms trade. I examined each of them once more—especially the preface, table of contents, and conclusions—to determine why each of these authors had written their book. I found in doing this that each of my predecessors had implicitly tried to explain how their treatment of the subject was an appropriate, necessary, and cumulative addition to the literature. A review of these books also represented a brief history of the arms trade and world politics, especially as the perceptions of trade patterns, purposes, and effects shifted over time. In general, these books were written in response to the policy issues salient during the period.

So it is with this book. Why a book on the international arms trade in 1992? One answer is that in the space of two years (while this book was being written), one of the two dominant suppliers in the international arms trade system, the Soviet Union, disintegrated. The preliminary evidence is that the successor republics, particularly Russia and the Ukraine, will in no way fill the gap. Not only does this mean a significant decline in arms trade but also a major change in the nature of that trade. Gone is that element of the system which promoted political/ideological rationales over military and economic. It is an occasion to reflect on earlier arms trade systems for clues as to how arms trade will work in the future.

Another major reason for a book now is the recent events in the Persian Gulf, which saw a developing country, Iraq, use legal and illegal means to import enough military capability not only to invade Kuwait, but also to threaten Israel and to defy the world's major powers. As a result of this war, a rare consensus has emerged on the negative consequences of an arms trade system that contributed in a major way to a war costing over $50 billion and nearly 100,000 casualties. The major powers and the United Nations are meeting to discuss this arms trade system, and over the next several years there will be no shortage of articles and books on this particular case and its aftermath. Although this book will contribute both directly and indirectly to the literature on this seminal event, the book's premise predates the Gulf War and offers a much broader set of conclusions.

There is a great deal of evidence that this arms transfer event occurred because the actors in the international arms trade system had paid insufficient attention to changes in the system. New situations—that is, change—and a lack of understanding of these situations are always reasons to write a book, and much of the arms trade literature reflects these motives. The very first major study of the post–World War II period, the 1971 book *The Arms Trade with the Third World* by the Stockholm International Peace Research Institute, declared the subject "understudied" and "surrounded by a good deal of sensationalism fortified by secrecy." One of the most recent comprehensive treatments of the arms trade was Louscher and Salomone's *Marketing Security Assistance* in 1987, whose purpose was a critical and empirically rigorous examination of the emerging "new marketplace." My book is also driven in part by the necessity to clarify further the recently perceived changes, most explicitly by emphasizing the shifts in international arms trade regimes and systems over the past fifty years.

My review of the books written on the international arms trade also reveals that most have been written at the nation-state level of analysis and are mainly descriptions and explanations of how specific arms-supplying nations make policies and decisions on individual arms transfers. Authors decry the lack of understanding at this level and attempt to explain the sometimes arcane nature of the policy-making process in the hope that policy outcomes will be "improved," "more rational," and so forth. When Iraq invaded Kuwait in August 1990, and the world asked how the former was able to amass such an imported arsenal, almost all of the explanations were at the personal level (for example, life histories of Saddam Hussein) or the national level (misguided U.S. foreign policy, lax export rules in Germany, and so on). This book seeks to move beyond the nation-state level, demonstrating, for example, that Iraq's acquisition of enough military capability to hold the major powers at bay is but one of many cases where the explanation for national behavior can be found at the international systemic level.

What, then, are the unique objectives of this book? First, it is an explicit attempt to connect international relations theory to a concrete phenomenon that is at the heart of the contemporary state system: the transfer of military capability for the purpose of securing and developing the national security of sovereign states. Enough arms transfer data exist to apply various theories to this important facet of world politics. A review of the existing literature reveals that arms trade research addresses bits and pieces of theory (systems theory, bureaucratic politics, dependency, and so forth). This book attempts to apply what has been learned about international politics at the systemic level to further understanding of the arms trade. In this book, four historical international arms trade systems are described and compared, providing the reader with insights as to how the structure and rules of a system can shape national and subnational arms transfer behavior. It also addresses the causes and implications of systemic transformation, using these conclusions to forecast what arms-trading systems will look like in the twenty-first century. It should also be remembered that this book is part of a series of texts to be used to further our understanding of world politics. Therefore, an important objective is to ensure that this study of the arms trade issue-area leads to a better understanding of the international system as a whole.

There is an explicit attempt in this book to use the systemic approach to reach those charged with making, analyzing, or evaluating policy involving arms transfers. Chapter 7 is designed to demonstrate how national governments and subnational actors respond to the regime and systemic shifts depicted in chapters 4, 5, and 6, as well as how systemic pressures contribute to explaining the arms transfer behavior of these actors in the system. And unlike other important phenomena (such as oil, trade, the environment), arms transfers are very much under the control of national governments. This places a particular urgency on the need to see how international environments shape national policy outputs.

A third major objective is to provide a book comprehensive enough to serve as a primer for those investigating the arms transfer phenomenon—a growing and diverse audience, given the Gulf War and the accompanying attention being paid to controlling and dealing with the arms trade. The goal of my book is to provide a treatment of the subject for both those familiar with the subject and those delving into the issue area for the first time. The book allows the reader to view arms trade at the level of the international system, sufficiently equipping him or her with enough knowledge to proceed with more in-depth investigation and study of specific aspects of the phenomenon at other levels of analysis. Chapters 1 and 2, which review the literature and data on the arms trade, contribute to this objective. Those readers more familiar with the field can move directly to chapters 3 and beyond, which contain substantive descriptions of the four arms transfer systems that have existed since 1930. In short, this book has been designed to be one of the first books pulled off the shelf when studying the arms trade.

Any book is going to reflect the experiences and biases of the author. My research and policy experiences in this field have greatly affected my approach to the subject, especially the treatment in this book. Since 1977, I have offered a seminar twice a year at my institution, the Naval Postgraduate School, entitled "Problems of Security Assistance and the International Arms Trade." It should not be surprising that I have organized this book as I teach that seminar: first discussing the literature, then issues of measurement and data, and then launching into the substantive material, starting with the international systems level of analysis. Most of my students (all of them are military officers from the three services) are being educated as area specialists, and many go on to positions involved with the real world of arms transfers. Some end up in Washington in the policy analysis business; others report to embassies where they are involved in observing and shaping the effects of arms exports from the

United States and other countries. The obvious result of this seminar is a policy focus. Just as importantly, however, I have stressed to my students the need to take a comprehensive and theoretical look at the phenomenon, lest they arrive at their post and be swallowed up by the minutiae, chaos, and policy biases of the practitioners.

I have also interacted extensively with U.S. arms industries, especially their representatives concerned with the marketing of products for export and interacting with the U.S. government. I have taken my students to defense plants for briefings on the arms trade from the industrial perspective. I have given two-day seminars to groups of industry representatives on the international arms trade system and the rules and regulations that govern the export of arms from the United States. Although I am sometimes taken aback by the lack of knowledge of these people, more often I am the recipient of insights often overlooked in most studies of the arms trade. Often the role of industry is a forgotten element in analysis of the arms trade, dismissed as either irrelevant or hopelessly corrupt and evil. I have tried in this book to take a more rational look at this role.

In 1978-79 I spent a year as special assistant to the chief of the Arms Transfer Branch of the United States Arms Control and Disarmament Agency. This was a year of unmeasurable value as far as understanding how policy and decisions are made regarding arms transfers. It was an unparalleled opportunity to verify international relations and foreign policy theories, especially in the context of bureaucratic politics. This was the year that the Carter administration was deeply involved in an attempt to control the international trade in arms. In December 1991, the General Assembly passed a landmark resolution by a vote of 150-0, establishing an international arms trade register. In January 1992, I was appointed as one of two consultants to the panel of national government experts charged with developing the procedures for the operation of that register. These two experiences deeply enhanced my knowledge of how the arms trade works, especially from the competing perspectives of U.S. and non-U.S. suppliers and the Third World recipient nations. It has given me a much more sober and balanced perspective of this phenomenon, which I hope is reflected in this volume.

Much of my research and writing on this subject has been related to the control of the arms trade. As a teacher of students who will analyze and employ arms exports as a foreign policy instrument, as well as students dedicated to eliminating the negative consequences of such exports, I have accepted the predominance and utility of this instrument. However, I have become particularly sensitive to the negative effects of the arms trade and continue to feel that these effects can and must be diminished. My practical experience and interaction with students who are from the practical policy world has given me a more realistic view of what can be done. But I make no apologies for my continuing concern that the transfer of arms creates significant problems for individual suppliers, recipients, and indeed the international system as a whole.

As indicated in this book, the future arms trade system may be much less benign in this regard. My book is dedicated to preparing us as policymakers and world citizens to cope with this future.

It is difficult to adequately acknowledge all those who have played a part in the development of the ideas and knowledge found in this, my first book. My friend and colleague Dan Caldwell has encouraged me for years to put my knowledge of arms trade into a book, and to him I am very grateful. Ed Kolodziej has been a source of intellectual inspiration for many years, and I thank him for his comments on a draft of the book and his many contributions to my knowledge over the years. Those of us academics in the United States who have continued to hammer away at the major questions dealing with the arms trade for the past 15 to 20 years are a small but persistent band—Stephanie Neuman, David Louscher, Mike Salomone, Michael Klare, Jo Husbands, Bob Harkavy, and Fred Pearson. I have enjoyed their company and I trust they find this book a worthy addition to the field.

I am thankful for the love and support of my wife, Martha, who has encouraged me throughout this and the many other projects I have been involved with. And finally, I would like to dedicate this book to my mother and father, Alice and Edwin Laurance, for their unfailing love, patience, and support throughout my journey from soldier to teacher and scholar.

# 1
# An Introduction to the Study of International Arms Transfers as a Concept and Issue Area

The study of international relations and foreign policy has taken several paths to better understanding the modern international system in terms of its actors, interactions, and effects. In taking a theoretical approach, one can logically develop propositions and models in order to make sense of the multitude of phenomena that make up the system. Another approach is to immerse oneself in the data of international relations, arriving inductively at empirical relationships. Yet another approach is more descriptive, patiently examining a particular actor, event, or phenomenon in the hope that a fuller understanding emerges. This book attempts a synthesis of these approaches by selecting for study one phenomenon—the international transfer of conventional arms—and subjecting it to empirical and theoretical analysis so as to illuminate and further clarify the working of the international system as a whole.

## Arms Transfers as an Important Issue Area

The success of such an approach depends first on selecting a phenomenon that is central to international relations. If a list were made of the key international events and movements of the twentieth century, how many of them would involve the transfer of conventional arms? Consider the following list:

**The Post-World War I Attempts at Disarmament.** One of the key suppositions of all of the debates of this era was the linkage between the acquisition of arms and the outbreak of conflict. The "merchants of death" theory that dominated this era held that it was an uncontrolled arms-trading system run by private firms willing to sell to both sides in a conflict that brought about World War I. Understanding the arms trade of this period takes one a long way toward understanding the central role of disarmament and pacifism during the interwar years, as well as the buildup of armaments associated with the outbreak of World War II.

**U.S. Entry into World War II.** One of the key decisions leading to the U.S. entry into the war was that regarding the so-called lend-lease agreements in which the United States leased major quantities of naval equipment to the United Kingdom. The debate within the United States over the negative consequences of this seminal arms transfer was far reaching and was to become typical of those that took place in the postwar period.

**Formation of Opposing East-West Blocs after World War II.** The contest for clients in the cold war centered on the use of the arms trade as an instrument of foreign policy. Since the United States and the Soviet Union rarely faced off against each other, proxy wars featuring combatants supplied with arms by the superpowers dominated the international security agenda. Acquiring arms from one or the other of the superpowers became primary evidence of a state's allegiance in the cold war.

**The Antiapartheid Campaign of the World Community.** By the 1960s, the world had normatively condemned South Africa for its racial system, a system that shaped African politics for many years after the continent's various states gained their independence. However, the one policy area chosen to put some teeth into this outrage was that of the arms trade. The success or failure of the opposition was measured by the success or failure of the arms embargo.

**The Vietnam War.** It should not be forgotten that despite the massive involvement of U.S. military forces in this conflict, it began and ended with a focus on the arms trade. Vietnam was one of the first tests of the Kennedy policy of sending arms and advisors to achieve U.S. objectives. The presence of over 500,000 U.S. troops by 1968 was, in effect, evidence of the failure of this instrument of foreign policy. Likewise, the inability of a massive influx of military aid after 1972 to create a South Vietnam capable of resisting attack from the north served as further evidence of the shortcoming of this tool. This arms transfer experience led directly into U.S. congressional efforts to pass major legislation significantly increasing congressional power on this issue.

**The 1973 Yom Kippur War.** The many accounts of this war focus on how the combatants employed their equipment (tactics, doctrine, and so on). However, it should be remembered that almost all of the equipment was imported, and in the final analysis, the two superpowers were very influential in using resupplies of armaments to affect the outcome of the conflict.

**The Camp David Accords.** Despite the deserved attention given to the principals involved in this historic event (Sadat, Begin, and Carter), a major key to its success lies with the massive increases in arms promised to both sides by the United States to ensure the perception if not reality of peace.

**The 1991 War in the Persian Gulf.** The focus on the $50 billion military campaign of the coalition forces during Operation Desert Storm ignores the fact that the entire campaign was necessary only because of an international arms trade system that allowed Iraq to acquire the capability to threaten Israel and other neighbors with attack by ballistic missiles with chemical, if not nuclear, warheads. A consensus has now emerged that changes must be made in the system so that such costly military campaigns are less necessary in the future.

In all of these events, which span several shifts in the international system, the use, limitation, or control of arms trade was central. Obviously, many political events and developments have had little to do with this phenomenon, and the events listed above turned on much more than their arms trade component. But the point to be made here—indeed, the rationale for this book—is that by studying this phenomenon empirically and theoretically, we will be rewarded with a better understanding of the larger whole: the international system and its effect on the policy choices of its actors.

What is meant by the terms *arms transfers* or *arms trade*? At its basic level, the arms trade concerns the acquisition and maintenance of national security by nation-states. In most cases, this security has been achieved through the acquisition of military capability. This capability comes in the form of military equipment and support, and it can be either indigenously produced or imported from another supplier (a commercial firm or a country) via international trade. Since the advent of modern weapons systems in the nineteenth century, a significant percentage of this capability has been acquired through trade, making such trade worthy of study.

Through the years, a host of terms have been used to describe this international phenomenon: *arms transfers, arms trade, military* and/or *security assistance, military cooperation,* and so forth. In the past twenty years, military capability has become even more difficult to conceptualize as a result of advances in technology, the growing importance of upgrading through components, the rise of coproduction, licensing and technology transfers, and the force-multiplier impact of command and control capabilities that are now transferred internationally. In order to minimize confusion, this book will refer to this phenomenon as *arms transfers* or *arms trade,* to signify that we are concerned with observable commodities that are traded in the international system for the purpose of enhancing the military capability or political power of the recipient nation.

As a previous volume in this series revealed, those researching international relations continue to debate which paradigms and theories are most useful in explaining the evolution of world politics.[1] This volume addresses that debate in chapters 2 and 3. But no matter how one views international relations theoretically, it cannot be denied that the production and transfer of arms have absorbed a major segment of the world's resources since the nineteenth century and deserve study on that basis alone. But the transfer or trade aspect of this phenomenon goes far beyond an important economic phenomenon in the same category as oil, technology, food, and other commodities that have been studied. Although it is true that trade in armaments is not immune to market forces, its potential to affect a nation's security has assured its treatment as a phenomenon uniquely relevant to world politics. No sooner had modern armaments manufacturers blossomed in the nineteenth century than governments began to control the exports of this and no other commodity. "The motivations were complex . . . in part on humanitarian grounds (to help eliminate the slave trade), in part on economic grounds (to preserve a lucrative market), in part on security grounds (to keep modern weapons out of the hands of potentially hostile natives)."[2] In one of the first major studies of the post–World War II period, Kemp and Sutton stated that "modern armaments cannot be equated with ordinary engineering exports; if they could, it would be proper for normal economic forces to determine the level of armaments in a given area. But everyone is aware that the sale and transfer of modern armaments has gone beyond the bounds of ordinary laissez-faire economics."[3]

The political salience of this commodity can be seen by observing how most Third World nations have responded negatively to suggestions on the part of the superpowers and other major arms suppliers that some sort of arms trade control regime be established. These countries jealously guard their right to defend themselves (and to attack their neighbors if necessary). As will be seen, up through the mid-1960s, the means to wage war was controlled by a handful of major suppliers. Acquisition of military capability and arms imports were synonymous. In their landmark 1972 study, Stanley and Pearton observed that "governments have become enmeshed in the arms trade because of the intense political significance that it has assumed. . . . The plain fact seems to be that arms transfers are rooted in the international system of sovereign states itself. . . . Governments have come to regard arms transfers as having an integral relationship with their own foreign policies."[4]

In addition to its role as a critical instrument of foreign policy, which inexorably ties it to international political forces, there are other aspects of this phenomenon that make it a valuable candidate for illumination of the international system as a whole. The arms trade has mirrored the rise of trade in high technology, an important environmental factor in the international system. Arms transfers have been greatly affected by the major economic developments in the international system: going off the gold standard, the oil price increases, the rise of debt in Third World countries, and so forth. Despite some significant measurement and data problems caused by the secretive nature of the trade, observers of this commodity have been able to generate enough empirical indicators to justify inquiry into the most critical of theoretical questions relating to change and transformation of the international system. The salience and visibility of this commodity has allowed us to know when the number of actors (suppliers and recipients) is increasing or decreasing, when the magnitude of military capability being transferred undergoes step-level changes, and when and how arms transfers impact on important systemic phenomena such as conflict, negotiations, economic development, dependence, and leverage.

The study of international arms transfers is also important because of the normative policy implications. The reality is that most of the death and destruction wreaked on the world, especially since the end of World War II, has been accomplished with armaments, logistics, and technical support imported by the warring nation-states and insurgent groups. The transfer of arms is not without negative consequences, and historically this aspect of the trade has prompted most of the study and research. Though this book has a theoretical and policy analysis focus, it is also clear that the normative questions cannot be ignored. This will be increasingly true if the structure of the international arms transfer system continues to evolve away from its postwar state-centric nature to one more closely resembling the merchants-of-death era of the early twentieth century. States are gradually losing control of arms transfers as military capability is increasingly defined in terms of technology and "black boxes" that are not only less visible in an empirical sense but also in terms of the normative reactions that result from negative political, economic, and military consequences.

## The Evolution and Sociology of Knowledge of the Study of International Arms Transfers

Although a case has been made that the study of the arms trade can lead to a deeper understanding of global politics writ large, the fact is the literature of the arms trade has been very uneven in terms of the historical periods, suppliers, recipients, rationales, and effects studied. At this point, a brief review of past research will be conducted to ascertain the salience of the arms trade phenomenon and to establish its connection to the larger issues of global politics. The review will also clarify the role of this book and its contribution to the literature as a whole.

### The Negative Effects of the Arms Trade

The first major conclusion about research on the arms trade is that it tends to occur in concert with perceived negative consequences. As World War I drew near, there was an outburst of studies that addressed the role of the arms trade in the arms races that were perceived as a major cause of the war. In the interwar years, as pacifism and disarmament began to play a major role on the world scene, the international arms trade again surfaced as a potential villain. (The interwar years are treated in depth in chapter 4.) The fact that several international conferences were held on the subject added grist to the mills of journalists and other observers of that period. When the League of Nations actually began to publish the *Statistical Year-Book of the Trade in Arms and*

*Ammunition* in 1924, it also forced a reaction from governments. Since the statistical yearbook was designed to focus on the negative aspects of the trade, this linkage was not surprising.

The next great outpouring of research on the arms trade did not occur until the late 1960s and early 1970s. The arms trade played no role in a World War II fought with indigenously produced weapons. After the war the international system was bipolar, with the United States and Soviet Union as the only significant arms suppliers. Furthermore, they were in the business of giving their clients surplus equipment. In short, this may have been a dangerous time in terms of threatened conflict between the superpowers, but it was a noncontroversial time in which both national and international public opinion saw little wrong with the arming of client states to prosecute a popular cold war.

By the late 1960s and early 1970s, however, the system had changed. The two blocs had begun to break down, western Europe had recovered and produced several new arms suppliers, and the major wars in Vietnam and the Middle East had begun to once again produce commentary and theorizing as to the negative effects of the arms trade. Some of the research was similar to that of the 1930s (that is, normative and focusing mainly on the aforementioned negative effects). However, for the first time, much of the research of this period was empirical and analytical. In either case, though, the occasion for the research was the perceived negative consequences among the concerned publics and policymakers.

The 1970s saw a series of policy initiatives by the U.S. Congress and executive branch to address these consequences. The Arms Export Control Act of 1976 and the Carter arms transfer restraint policy created the impression that action was being taken to address the negative effects of the growing arms trade. As a result, the literature on the subject declined accordingly. As the U.S. public mood became more anti-Soviet in the late 1970s, little attention was paid to the arms trade. Ironically, a series of shifts in the international system, including the rise of even more suppliers to compete with the superpowers and a sharp decline in the ability of Third World states to pay for arms, resulted in an overall decline in the arms trade and its study. It was not until the late 1980s that the arms trade once again began to surface as a research topic as a result of the perceived negative consequences of the increasing proliferation of ballistic missiles and the ability to make chemical warheads.[5]

As we enter the 1990s, the study of the arms trade is once again in full swing, mainly because of the Gulf War and the perceived negative consequences of an unrestricted arms trade that very suddenly received empirical verification in the case of Iraq's acquisition of sophisticated and destabilizing military capability. Much of the current literature is reminiscent of both the 1930s and the 1970s writings referred to in the above paragraphs. The work on ballistic missile proliferation and its consequences has accelerated. Editorials and op-ed pieces across a wide variety of media calling for arms trade control appear almost weekly, many with titles echoing the themes of previous periods: "Stop Arming the Third World," "Avoiding New World Disorder," "Disarming the Gulf," "Here We Go Again—Arming the Gulf," "Arms and the Poor," "How to Limit Everybody's Missiles," and so on. In March 1991, a monthly newsletter published by the Federation of American Scientists, *The Arms Sales Monitor,* began to track congressional and other activity on the issue. Even the cartoonists[6] and humorists are back in action. As only one indicator, Art Buchwald is once again producing columns on the more negative and ironic aspects of the arms trade—three columns in ten months.[7]

*Research as a Function of Arms Transfer Policy Initiatives and Events*

**The 1930s.** Not surprisingly, the study of the international arms trade usually coincides with the development of various national and international public policy initiatives that surface as a result of perceived negative consequences, especially in the United States. The outpouring of research in the 1930s was due to several international conferences and the development of the statistical yearbooks by the League of Nations. International conferences (such as the 1925 Conference for the Control of International Traffic in Arms, Munitions and Implements of War) required the participating nations to prepare policy papers in support of their positions. The meetings were public, promoting analysis of these positions by journalists and other analysts. Within the United States, this was also the time of the Nye committee hearings examining the entire set of arms trade issues from the U.S. perspective (see chapter 4).

**The 1970s.** In the 1970s, the U.S. Congress began to investigate the increased and changing nature of the international arms trade. This attention resulted in an increase in support of arms trade research, not only by the executive branch in response to these initiatives, but also by the foundations that generally supported the more critical work of the independent universities and think tanks. Since Congress lacked any significant in-house capability to study these issues, knowledge for use in the policy deliberations was typically generated by hearings and commissioned reports on the key arms transfer issues.[8] Conferences on the arms trade began to occur frequently, generating papers, books, and conference proceedings.[9]

Once the major legislation (the Arms Export Control Act of 1976) had been written and the Carter administration had made concern for the negative consequences of the arms trade a major pillar of its foreign policy,[10] research and publications on the subject tended to decrease. From the perspective of a participant in the bureaucratic politics of arms transfers,[11] policy analysis began to dwindle once the issue began to receive less attention from the president. For example, Carter's policy had included as a basic element discussions with the USSR regarding the reduction and control of the international arms trade; these talks collapsed in December 1978.[12] Prior to their collapse, U.S. Arms Control and Disarmament Agency (ACDA) Director Paul Warnke resigned. He had been very close to the president and as a result was very influential in the bureaucratic struggles that surrounded the Carter arms transfer policy. His leaving was an indicator that the issue was moving down in priority.

Another component of the arms transfer policy was a ceiling on U.S. arms exports. In the first two years of the Carter administration, this very controversial policy initiative sparked a huge outpouring of analysis and commentary.[13] By the summer of 1979, little attention was being paid to its utility as a method of controlling arms exports. By the end of 1979, with the fall of the shah of Iran and the subsequent taking of U.S. hostages, as well as the Soviet invasion of Afghanistan, the arms transfer policy had all but disappeared from the agenda, both in the executive branch and in Congress. What remained to be accomplished in the way of analysis were the predictable postmortems of a perceived failure of the Carter policy and the policy prescriptions for a new (and "more realistic") approach of the next administration, be it Democratic or Republican.[14]

**The 1980s.** With the advent of the Reagan administration in 1981 and its public mandate to increase not only domestic defense spending but also arms exports to compete with the Soviet Union, policy initiatives regarding arms transfers diminished, as did research on the subject. In a sense, the literature became reminiscent of the 1950s and 1960s, when most policy research was geared to fine-tuning the bipartisan cold war military assistance policy of the United States. In the U.S. Congress, there were pockets of critics of the international arms trade, but their ideas rarely got beyond testimony at hearings and in the *Congressional Record*.

Andrew Pierre's *The Global Politics of Arms Sales*,[15] published in 1982, was the last of the comprehensive treatments of the international arms trade spawned by the arms transfer policy debates that marked the 1970s. Six years passed until Catrina's 1988 book *Arms Transfers and Dependence*,[16] which is an excellent summary of the field as well as an in-depth treatment of propositions related to arms transfers and influence, leverage, and dependence. Only the Stockholm International Peace Research Institute (SIPRI) continued consistently to research and publish data and analysis on the topic during the 1980s. SIPRI's yearbooks each have contained a major chapter on the international arms trade since their inception in 1968. Since 1986, SIPRI has published three major volumes that qualify as definitive works: *Arms Production in the Third World* was published in 1986 and provides a theoretical overview of this phenomenon and individual-country studies through 1984. *Arms Transfers to the Third World: 1971–85* was published in 1987 and contains a comprehensive set of data, as well as an analytical focus that describes and analyzes the evolution of the international arms trade system, the rationales and styles of major suppliers, and a brief treatment of the interaction of supply and demand. *Arms Transfer Limitations and Third World Security,* published in 1988, is an excellent summary of past efforts to control the arms trade and of some suggested alternatives.

Most of the non-SIPRI research being conducted was prompted by the new phenomena that had emerged in the international arms trade system. These topics included offsets, second-tier suppliers, and the growing tendency toward the internationalization of military production. But compared to previous periods when arms trade research occupied center stage, relatively little was being done to monitor and evaluate the changes taking place.

This lack of research in the 1980s was primarily a function of the lack of policy initiatives on the issues. From the advocates of increased arms trade, very little in the way of policy initiatives was needed, since the consensus that arms sales had declined worldwide was well documented. Meanwhile, most suppliers (commercial and governmental) were trying to sell as much as possible. No doubt the arms industries of the world continued to conduct their traditional market studies, but these rarely surface as publicly available research.

From the control side of the issue, several things contributed to a lack of those types of policy initiatives that spawn research. First, the empirical reality was that, in the aggregate, global arms sales were diminishing. If the objective of the control efforts of the 1970s was "less," then success had been achieved, and little new action was called for. A second possible reason for the lack of arms control activity was the Carter experience. Here was a high-level effort that attempted to reduce the negative consequences of the arms trade, and it had failed; many saw this as the last, best hope to gain control of the problem. The developments of the 1980s had done little to boost the morale of these advocates, given the proliferation of many new and uninhibited suppliers in the system and events such as the Iran-Iraq war, which was fought mainly with imported military capability. A pessimism had set in that the arms trade was inevitable, and there was little impetus to do anything about its negative effects.

A third factor in this lack of research was the evolution of U.S.-Soviet relations, particularly the fact that the major war of this period (Iran-Iraq) was not of major superpower concern. Particularly important here is the fact that the United States didn't supply arms to either combatant,[17] and the USSR supplied arms mainly to the side fighting a U.S. enemy, Iran. Missing this time around was the "U.S. as arms supplier" label that was the focus of much previous research activity. Of course, all of this changed when Iraq invaded Kuwait: the former "client" of both the East and West had turned on its benefactors.

This is not to say that U.S. policy was not concerned about Soviet arms transfers. In fact, the Reagan arms transfer policy was formed with a specific anti-Soviet flavor, which justified a loosening of national export controls in order to counter Soviet arms transfers to countries such as Ethiopia, Angola, Afghanistan, Syria, and Nicaragua. The point is, however, that none of this was controversial enough to generate policy initiatives that either required or prompted the flurry of research that has marked previous periods. An example that makes this point is the Missile Technology Control Regime (MTCR), an agreement between seven states (the United States, Canada, Germany, France, Italy, Japan and the United Kingdom) that was signed in 1987. The seven countries agreed to a "public set of common guidelines and a common annex of items to be controlled, with the focus on delivery systems rather than nuclear warheads."[18] The initiative for this agreement dates back to the early 1980s but did not appear on the public agenda until the actual signing. Even then, compared to previous periods, it produced little comment or analysis. It was not until 1989, when it became public that the Iraqis and others were developing advanced ballistic missiles and chemical weapons, that the MTCR began to be examined in the literature.

In the final analysis, the transfer of conventional military capability in the 1980s was not seen as a problem worthy of much research. Certainly changes were taking place, but they were incremental in nature and did not involve the level of system transformation that had marked previous efforts. Lurking in the background was the feeling that the proliferation of conventional weapons, no matter how advanced and numerous, simply did not represent the level of threat posed by the proliferation of nuclear weapons. No scenarios could be imagined that could match the level of crisis and disaster envisioned by the explosion of nuclear weapons by Third World states.

**The 1990s.** As the decade of the 1990s begins, we again see research increasing in response to policy initiatives. A sudden surge in the supply of ballistic missiles to potentially hostile states, and their use by Iraq, has the major powers of the world very concerned and generating policy initiatives in a variety of forums. Within the United States, both congressional and executive-branch policy alternatives are being debated. At the July 1991 meeting of the G7 countries (the United States, Japan, Germany, France, the United Kingdom, Italy and Canada), the final communique included a call for an arms trade register and a declaration of the need to address ballistic missile proliferation. Within the United Nations, the various resolutions involving the restriction of arms exports to Iraq have focused that organization more generally on the arms trade issue. The five permanent members of the UN Security Council met in July 1991 to begin a process of controlling the arms trade, especially missiles with warheads capable of mass destruction. Likewise, the efforts to settle the conflicts in the Middle East will likely have an arms trade control component. In December 1991 the General Assembly passed the Transparency in Armaments resolution, by a vote of 150-0 (two abstentions). It established an international arms trade register and an eighteen-country panel of experts to develop procedures for its effective operation by July 1992. As policy speeches and efforts to address this issue in international forums proliferate, so do the concomitant conferences and studies, as well as the research money needed for their support.

## Arms Transfers and the Behavioral Revolution in Political Science

Like the larger field of international relations, research on the international arms trade reflects trends in the development of political science and other disciplines, especially in the United States. For example, the legalistic and moralistic orientation of political science in the interwar years produced a literature heavily laden with these overtones. Harkavy, in his major review of the interwar literature, characterized this outpouring of writing on the arms trade as of the "muckraking or exposé variety."[19] Bibliographies of this period are also replete with references to numerous legal documents of the various international conferences that were held. Behavioralism, levels of analysis, and statistical studies aimed at verifying empirical realities were unknown. Therefore, one of the reasons for the major outpouring of research on the subject from the late 1960s to the mid-1970s was the changes that had taken place in the study of political science in the United States.

The first truly systemic research on the arms trade did not take place until well into the 1960s, in the wake of Singer's 1961 seminal piece on levels of analysis.[20] The first locus of such research was the Center for International Studies at the Massachusetts Institute of Technology (MIT). A group of researchers led by Lincoln Bloomfield and Amelia Leiss developed a project on the control of local conflict into the more specific assessment of the role of arms transfers in conflicts. The result was a computerized data base of all arms transfers to and from a fifty-two-country sample and a six-volume study that was summarized in the first volume, *Arms Transfers to Less Developed Countries*.[21] The major contribution of this research was its scientific and analytical rigor. Operational definitions were developed, actors categorized as to "styles," and hypotheses tested as to the effects of arms transfers.

At the same time, another research effort was commencing at SIPRI, which had just been created. The motivation for this research was different than that of the MIT project. SIPRI was addressing the arms trade in an effort to resurrect the disarmament efforts of the 1930s, especially the creation of arms trade registers. In 1971, SIPRI produced a 910-page book entitled *The Arms Trade with the Third World*,[22] which became the major work on the subject. Though the motive for the research was clearly one of "finding out what measures, if any, could be taken to limit this part of the global arms race," the research was as thorough, analytical, and comprehensive as the MIT studies. In the book's own words, "SIPRI decided to study all aspects of the arms trade between developed and underdeveloped countries. What were the facts about the arms trade? . . . What was the role of the arms trade in the relationship between exporting and importing countries? What were the consequences for third world conflicts, and for internal developments in recipient countries? What proposals have been made to regulate the arms trade?"[23]

This project also produced a data base that, unlike the MIT effort, has continued to the present day. The initial project developed a methodology for operationalizing the magnitude of a particular transfer of major weapons systems, based on economic and military factors. In addition, it produced arms trade registers that described the individual deals for each country, and SIPRI continues to do so each year in its annual Yearbooks. *The Arms Trade with the Third World* also included a major effort at the national level of analysis, but its systemic contributions greatly influenced the study of the arms trade. The book documented the international arms trade system as it existed in 1970, and it became the major source for concepts, propositions, and data.[24]

There were other books published at this time that contributed to the general knowledge of the phenomenon, but they tended to be either descriptive or journalistic.[25] In 1972, an excellent article appeared in the *Journal of Peace Research* that reviewed and debated this recent spate of research.[26] In 1975, however, the first truly systemic-level study was produced by Robert Harkavy in *The Arms Trade and International Systems*.[27] Harkavy had the benefit of the MIT and SIPRI studies, but his was a major advance for several reasons. First, he formally integrated the concepts of the growing literature of international systems. Secondly, he linked the arms trade system to the larger international system as a whole, showing how changes in supplier-recipient patterns, for example, are related to larger systemic concepts such as ideology and technology. Third, and most important, he focused on system *change,* describing and comparing the arms trade systems of the 1930s and of the 1945–70 period. In doing so he provided a methodological approach that can be used to explain further shifts in the system, as well as for forecasting future evolutions in the system. Harkavy's work will be used extensively in chapter 4 of this book.

## Questions, Propositions, and Puzzles

As a final step in this analysis of knowledge of the international arms trade, I will address those questions, propositions, and puzzles that have emerged to guide research on this issue area. Kuhn's classic work *The Structure of Scientific Revolutions*[28] remains very useful in describing exactly what makes up a "science" or field of study.[29] By Kuhn's definition, a *paradigm* is a pattern or framework that gives organization and direction to a given area of scientific investigation. A paradigm must first have concepts that can be used to devise propositions and to provide focus for the generation of the data used in empirical investigations. Secondly, it must have a theoretical element, a set of logically connected propositions that can be empirically tested. A third element is a set of rules of interpretation; this element includes commitment to a preferred set of measuring instruments and resulting data that are used to describe the observable phenomena that make up the field of study (see chapter 2). To qualify as a paradigm, these rules of interpretation must reach some level of acceptance among those investigating the phenomena. A fourth element of a paradigm is the identification of those puzzles that are worth solving and for which there is some assurance that a solution exists.

In 1977, Laurance and Sherwin reviewed all of the research on arms transfers that had taken place since the resurgence of the effort in 1970 to include published books and articles, panels at academic meetings, policy studies done for the U.S. government, congressional reports, and the several major conferences held on the subject. The propositions that emerged are listed below in the form of questions.[30]

## DESCRIPTIVE

1. What are the recipient countries' demands for conventional arms?
2. What are they contracting for, in terms of numbers, types, mode of delivery, and financial arrangements?
3. What are the delivery patterns (numbers and types)?
4. What is the monetary value of these transfers?
5. What is the military capability of the recipient country before and after a specific arms transfer?

## IMPACT ON SUPPLIER

1. Do various types and levels of arms transfers result in political influence and leverage for suppliers?
2. How do various types and levels of arms transfers affect suppliers' prestige, at home and abroad?
3. What economic benefits accrue to suppliers as a result of arms transfers?
4. How do arms transfers compare with other policy instruments in accomplishing foreign policy objectives?

## IMPACT ON RECIPIENT

1. How do certain types and levels of arms transfers affect local military balances?
2. Do arms transfers enhance or inhibit the internal stability of recipient countries?
3. What effects do arms transfers have on the economic development of recipient countries?

A similar inquiry was made in 1979 by Brzoska in a seminal piece entitled "Arms Transfer Data Sources." Before focusing on the data sources, he first asked the question, "What data do we want?" This produced what he termed "problem clusters" that "represent different groups of questions that are generally answered with arms transfer data."[31] These clusters were as follows:

1. *Arms industry*—arms exports as outlets for arms production, mainly in industrialized countries, and thereby connected with employment, growth, and balance of payments problems in industrialized countries. This cluster includes political as well as economic problems.
2. *Recipient's burden*—arms imports as a burden on the present and future resources of recipient countries with regard to foreign exchange, manpower, infrastructure, and industrialization patterns. Here again, mainly economic and political questions are addressed.
3. *Military use value*—arms exports as transfers of potential military use value, depending on the characteristics of the hardware, the environment in the recipient country, and surrounding transfers of support equipment, services, construction material, and the like. While military assessments are necessary at the outset, questions primarily of foreign policy and internal policies are asked here.
4. *Dependence*—arms transfers as a means of extending influence, either by direct leverage through abrupt changes in transfer policies or indirectly through the need for repayment. This also includes questions of political (and moral) compliance. This cluster refers to supplier-client relations on the political (and, by extension, economic) level.

In 1987, as part of a three-year project on the international arms trade, I developed a set of major issues and topics that were then presented to potential participants in the project at various locations. In essence, it was an opportunity to define the field of study in terms of the propositions that defined the paradigm and the gaps in the literature, at least as I perceived them.

1. The Arms Trade as an International System: Descriptions and Models
2. The International Arms Trade as an Economic Phenomenon
3. The Politics of Arms Transfers: Supplier Influence and Recipient Leverage
4. The Role and Dynamics of Technology in the Arms Trade
5. New Arms Suppliers: Their Motivations, Styles, and Impact
6. Arms Transfers as Tools of Conflict Control
7. The Impact of Regional Factors on Arms Trade Patterns

As this book goes to press, major changes are occurring in the international arms trade system; these are addressed in detail later in the book. This restructuring of the international system may produce new suppliers of arms as well as new motives, and the arms transferred may look very different than those that dominated the previous systems. Although these changes are no doubt producing additional questions and propositions, a basic premise of this book is that on this dimension the basic paradigm remains intact. The questions and puzzles of the past still serve as excellent guides to research. For example, those evaluating the case of Iraq and the arms transfer dimension of the recent Gulf War are served well by the above list of questions and propositions. This may not be as true for the generation and measurement of data needed to answer these questions; it is to this problem that I now turn.

# 2
# The Conceptualization and Measurement of International Arms Transfers

In chapter 1, the reader was introduced to the research and literature that has evolved around the study of an important concept in international relations, the international transfer of arms. That chapter concluded by outlining those questions and propositions that have dominated the field of study for the past twenty years. In chapter 2, I will look at the concept in terms of the analytical methods and data that have been used to address these propositions. Although very few subfields of political science meet all of Kuhn's criteria, especially when compared to the more developed "hard sciences," fields of study in the social sciences do approach paradigmatic quality as scholars and analysts work within a defined framework. The study of conventional arms transfers is no exception, and in this chapter, I will look in more detail at another element of this framework: the operationalization and measurement of the arms transfer concept.

## The Arms Transfer as the Basic Unit of Analysis

### Basic Dimensions of the Concept

Before we can begin to use the propositions and puzzles outlined in chapter 1 we must first ask, of what is an arms transfer composed? How can we know one when we see one? The validity of data (that is, agreement on the rules of interpretation) is a measure of how well one has operationalized the arms transfer concept, the assigning of a value to a case on a specific dimension. To illustrate the variety of definitions and dimensions of the concept, we can take the basic unit of analysis used to generate all data—the single case or arms deal—and briefly list its components. Most who have studied arms transfers since the 1930s will agree that the following composite list adequately defines the dimensions of the concept.

- *Dollar value*—How much did the recipient pay? What was the equipment worth, on the market and to the supplier? Dollar costs were one of the first indicators used to describe the phenomenon, even before the behavioral revolution, and have remained of primary interest.

- *Number of equipment units transferred*—This element of an arms deal has always been important, since variation in quantity is naturally perceived to be related to variation in effect.
- *Type and model of equipment*—The analyst will want to know this about the case, since other key dimensions will flow from this basic data point. Dollar values tend to vary with this information, as well as with military capability. If it is a new model, it may have political and prestige effects as well as military.
- *Modernity*—The equipment may be new, secondhand, or refurbished.
- *Military utility*—The equipment making up the deal will vary according to payload, speed, accuracy, and other characteristics that are aggregated to produce a military utility for the user.
- *Logistics and support services*—Often the services and materiel accompanying the transfer can be critical, especially if one is addressing comparative military capability. A tank with no spare parts or little ammunition is of less military value than one fully equipped on these dimensions.
- *Training accompanying the transfer*—Training can have military and political effects and is an important component of a deal. It can be a surrogate measure for the relative abilities of recipient country personnel to operate complex equipment.
- *Mode of payment*—There is a clear spectrum of payment modes, each with significant implications. One end of this spectrum is a pure grant or gift, followed by credits or loans, government-to-government cash, private industry-to-government cash, barter, and offsets. This dimension has significant implications for systemic transformations and the role of the arms trade in political relationships, as will be seen in the chapters to come.
- *Mode of production*—Another typology of values presents itself on this dimension. A deal could involve the transfer of an item "off-the-shelf," or it could involve any of a variety of arrangements such as licensed production, coproduction of specific parts, coproduction of the entire item, or codevelopment of the product with the supplier country. This dimension is particularly important for measuring the effect arms transfers are having on a recipient's indigenous production capability.
- *Delivery stage*—There are significant differences between an agreement and a delivery. However, there are other stages in the delivery process that also have systemic effects. In effect, the delivery stage runs from rumors (often promulgated by the recipient or supplier specifically for the purpose of testing the reaction) through negotiations, agreements, formal orders, and delivery to the various steps the recipient goes through to prepare the equipment to achieve its maximum military utility. On the military

18 • *The International Arms Trade*

dimension, a piece of equipment has a value even at the rumor stage if an opponent is forced to respond in some way. Conversely, only when operator efficiency is such that the equipment can be fully utilized can the military capability be said to have been truly "delivered."

## *An Example: The Recent Arming of Iraq*

To clarify how these dimensions can shape arms transfer events and the perceptions that emerge from them, it is useful to look at the recent case of Iraq's acquisition of significant military capability via imports. The *dollar value* dimension was critical in assessing this case. It is now known that Iraq's desperate financial condition in the late spring of 1990 was a major factor in pushing it toward invading Kuwait. Much of Iraq's massive debt had been accumulated by the arms expenditures of the Iran-Iraq war and in subsequent deals; therefore, how much Iraq paid for its various weapons systems is directly related to the conflict itself. And it is certainly related to the aftermath, as it is now known that many of Iraq's suppliers are left being owed significant amounts of money.[1]

On the other hand, using only dollar value would have seriously misled analysts as to the usable military capability being acquired. Although money and capability may be related for systems such as tanks and fighter aircraft (see the further discussion in this chapter), the financial dimension is much less useful when it comes to assessing Iraq's ability to upgrade its Scud missiles. A brief look at the items and technology acquired from Germany and the United States for this upgrade reveals that the cost was minimal, but the effects were crucial in explaining the conflict.[2] There was a major shift in the response of the major powers after Saddam Hussein threatened Israel with a chemical missile attack in the spring of 1990, a shift related less to monetary values of arms imports and more to specific technologies.

One of the key conclusions of the various postmortems of Operation Desert Storm was that the coalition forces overestimated the actual numbers of tanks and armored vehicles in the theater of operations, reminding us that the *number* of pieces of equipment transferred can also be crucial to any analysis. This was also true of the number of Scuds and their mobile launchers. Recent assessments of the ballistic missile proliferation threat also point out that these systems *may* be dangerous to the major powers, but that the threat turns very much on the quantity transferred.[3]

The implications of misusing the *type and model* dimension of an arms transfer was also illustrated during this event. The media became an instant resource for diagrams and pictures of the weapons systems being used in the war. Rarely were Iraq's ballistic missiles referred to by their proper name, al-Hussein; most often, they were referred to as Scuds. Not only are there three different models of Scuds, but in this case the al-Hussein was a Scud that had been upgraded in range using imported technology. No doubt some government specialists knew the difference, but it is much less certain that those actually determining policy responses did. If the missiles really had been Scuds from the USSR, the USSR could have been enlisted to help with the threat assessment and perhaps involved more in the resolution of the conflict. Since this was not the case, an entirely different response was in order, and policymakers found themselves seeking out West European and U.S. companies that had aided in the upgrade. The dimension of *modernity* also came into play in the case of the Scud/al-Hussein confusion. The age or shelf life of a missile can determine its effectiveness.

The *military utility* of the Scud-B missiles that Iraq acquired from the USSR in the 1980s was greatly affected by the range of the missile. Without alteration, their range was 190 miles. When upgraded (to al-Hussein), the range increased to 375 miles,[4] thus allowing Saddam Hussein to threaten Israel and change the very nature of the event. Perhaps a better example is the widely varying assessments of the effect of chemical warheads for these missiles; assuming that these warheads had been tested and installed, what was their military utility?

*Logistics and support services,* although not normally considered part of an arms transfer, can be crucial in assessing effects of the trade. In the case of the war with Iraq, it turned out that among the most critical pieces of equipment imported by Iraq were the bunkers to house the command centers and frontline troops. None of this trade ever appeared in the various compilations of arms trade data to be discussed later in this chapter. Trucks are another example of an item that is rarely monitored with any regularity, yet can be crucial if the real task is to evaluate a country's military capability as a result of the arms trade. Israel, for example, significantly multiplies its tank mobility by the acquisition of many flatbed trucks ("low boys") in the event that tanks have to be shifted quickly from one theater to another (for example, from Syria to Egypt). These strategically important vehicles do not appear in the order of battle as compiled annually in *The Military Balance.*[5]

Perhaps one of the most important dimensions of an arms transfer is the *training* and *technical assistance* that accompanies the transfer. In the fall of 1990, many Soviet advisers remained inside Iraq. What was their role? Were they needed to actually operate some of the weapons systems most threatening to the coalition forces? Although this story has yet to be told in its entirety, suffice it to say that it was a dimension of the problem for which very little information was available to aid the policymakers as they sought to fashion a policy response. Arms deals can be classified as to the amount of training and technical assistance required, but rarely is this the case. Furthermore, an identical inventory of weapons systems (say 100 Scud-B ballistic missiles) can take on a vastly different value depending on the capabilities of the recipient forces to operate them. Can the major powers assume that the capabilities displayed by the Iraqi missile forces in the Gulf War will be the same for the Syrians, who possess similar weapons?

The *mode of payment* also became a critical dimension in determining policy responses to Iraq's imported arsenal. As previously mentioned, much of the arsenal purchased in the latter stages of the Iran-Iraq war and the buildup to the Gulf War was done so with money borrowed from Saudi Arabia and Kuwait. When the Iran-Iraq war was over, these two creditors wanted their money, and Saddam Hussein wouldn't or couldn't satisfy their demands. Would the situation have been different if Iraq had been able to buy this military capability with cash? How much influence did the USSR have over the use of its equipment, especially compared to previous times when grant aid was the mode of payment?

The *mode of production* also came into play during this event. It became important, for example, to know if the missiles being fired on Israel and Saudi Arabia consisted of parts manufactured in the USSR (the al-Hussein was, in essence, two Scud-Bs linked together) or those manufactured in Iraq, with the latter items assumed to be less capable. Monitors of the arms trade were able to approximate how many mobile launchers Iraq had imported from the USSR, but were less certain as to how many Swedish trucks had been converted by Iraq for this mission.

Finally, the *delivery stage* of Iraq's ordered equipment became a crucial dimension. Once an embargo was installed in September 1990, there were many rumors as to specific items that had leaked through and forced a policy response on the part of the coalition. Exactly when the French, for example, stopped supplying software upgrades for Iraq's French surface-to-air missiles became a crucial piece of data for coalition military planners.

These, then, are the components of the international arms transfer. The above illustration was intended to demonstrate that, depending on the question being asked, various dimensions can be used. Scholars researching this phenomenon vary significantly in their emphasis on one or the other of these components, most often in relation to the propositions they wish to address. I now turn to a history and description of the data that have been generated based on the components detailed above.

## Generating Reliable and Valid Arms Transfer Data: Does It Make a Difference?

The preface and first chapter put forth several objectives of this book, one being to use the concept of international arms transfers to describe and explain how the larger and more general international system operates and changes. One of the justifications given for this effort was that the data were rich enough to lend empirical weight to the theoretical effort. Having briefly defined a consensus list of the components of the basic concept, this chapter now turns to exploring how these concepts have been measured, a necessary step prior to proceeding with any analysis of the arms transfer phenomenon and its effects. I will review the data that have been generated and used by scholars and the policy analysis community, assess their validity and reliability, discuss the importance of selecting appropriate data for the variety of existing propositions, and briefly discuss some of the methodological challenges in using the arms transfer concept. This is intended as an overview of the significant debate and critique of the past fifteen years[6] and as a necessary precursor to the rest of the chapters in the book.

Does the generation of valid and reliable data matter? A recent *New York Times* article reported that the U.S. government's system of gathering economic statistics is eroding to the point where it is affecting policy outcomes. A combination of budget cuts and deregulation is reducing the frequency of data collection and efforts to improve gauges that professionals think imprecise—foreign trade, for example. One part of the problem is the reliability and accuracy of data, with cuts in data collection affecting sampling techniques. Another aspect of the problem concerns validity, with statistics agencies being forced to

> skimp on research to make data reflect broader changes in the way the economy functions. Economists are debating whether important statistics like productivity, savings and the GNP really measure what is going on in the economy. . . . Lest anyone think this is a dusty debate for economists and bureaucrats, the truth is that these numbers affect the lives of all Americans. Social Security payments and some wages are tied to the official inflation rate . . . [and] the financial markets increasingly jump up or down, making or losing millions for investors with each new economic report.[7]

This phenomenon of measurement having an impact on reality is no less applicable to the arms transfer phenomenon. For example, the data-collection norms of the postwar period focused on nation-to-nation transfers of readily identifiable commodities: fighter aircraft, tanks, artillery, missiles, ships, and so forth. Furthermore, there was a consensus as to the effect that these systems had on deterrence and warfare. Despite the natural secrecy and sensitivity of governments, overall trends could be discovered by the intelligence agencies of the two superpowers. As the bipolar system began to break down, however, monitoring arms transfers became more problematic, and how these commodities were counted and valued became critical to the actual operation of the system and to national policy-making.

As one example, the production and exporting of prestigious and militarily capable weapons systems—for example, fighter aircraft—tell a great deal about the international security system. The production and export of fighter aircraft by single nations began to decline in the 1970s in the face of multinational production schemes. The Tornado aircraft is a good example. It was designed and built by the United Kingdom, West Germany, and Italy for use by the forces of these three countries. Systemic pressures to export, given the rising costs and shrinking domestic market, soon saw this aircraft being marketed to other countries.[8] This created two problems for analysts. At the international system level, how is one to calculate "market shares"? Technically, this aircraft is built by a multinational organization (Panavia), with a board of directors from all three countries. If seventy-two of these aircraft are sold to Saudi Arabia, which country sold them? Do we divide the dollar value up according to some percentage known only to the board of directors? This is not unimportant if we are to track systemic change and the rise and fall of supplier influence in the system.

At the national level of analysis, another set of problems arises if we are to accurately describe and explain the policies and decision-making processes in the major arms supplying countries. Germany (formerly West Germany) has become an increasingly active and influential member of the international arms trading system. This has occurred despite a declaratory policy that is indeed restrictive: arms are not to be exported to "areas of tension."[9] Despite producing one of the most effective main battle tanks, the Leopard series, Germany has been unable to sell it to Saudi Arabia because of perceived domestic pressure. And yet, in one of the most important and hotly contested arms deals in recent years, Saudi Arabia rejected purchasing additional U.S. F-15s in favor of seventy-two Tornados.[10] Throughout the aftermath of this sale, it was only the United Kingdom who was listed as the supplier, despite a handsome profit for Italy and West Germany. West Germany had changed its role in the system in real terms, but not in terms that could be detected by traditional counting norms, which focused on national government-level outputs. In this case, it was not so much secrecy that obscured reality as it was the methods by which analysts track the international trading system and national-level policy-making.

Another example is taken from the Carter administration, where arms export control was an important component of foreign policy. The policy was based on the assumption that the United States was the leading exporter of arms, which justified initial unilateral restraint by the United States while an international restraint regime was negotiated with other suppliers and recipients. No sooner had the policy been announced in May 1977 than data began to surface demonstrating that the Soviet Union, not the United States, was the leading exporter of arms. Some of these conflicting data were put forth by opponents of the policy, but most were simply products of bureaucracies that counted and valuated arms using different measurement techniques and assumptions. The CIA put out a draft of a document that, instead of dollar values alone, utilized actual transfers of equipment types; it showed the Soviet Union leading the United States in most categories. It took almost a year for this document to be issued, since it was primary evidence that one of the basic tenets of the arms transfer restraint policy was in dispute.[11]

Analysts of the international arms trade are in a familiar dilemma. Without some attention to theory, particularly regarding the evolution of the international arms trade system and the national response to it, analyses will tend to be at best eclectic, and at worse biased in the direction of the policy prescriptions of the writer. On the other hand, empirical evidence of this phenomenon is problematic, as soon will be shown in this chapter. What can be done? Louscher and Salomone, in their stinging critique of previous work, cite the normal response to this problem: "Unfortunately, many scholars are inclined to accept whatever data are available and conclude that their availability, especially if in some comparative and quantitative format, makes them useful for answering the questions raised above. Any doubts about the validity of the data are usually quickly dismissed by citing in a footnote one or two scholars who have expressed grave concerns about the uses of arms transfer data."[12]

One solution is to collect systematically data that are an improvement over what is now available. Kolodziej, at the end of his major treatise on French arms production and transfers, expressed his frustration with the current situation: "There is far more material available in the open record, particularly in open societies, than has been systematically reviewed and exploited by current research efforts. Gathering these data is a prerequisite for theory building, either in explaining global arms production and transfer or their implications for international stability, and for moderating hostilities and terminating them when they erupt."[13]

But, as Kolodziej and others have concluded by their continual use of the data that are available, analysis of this important international phenomenon cannot wait until such an effort is a reality. Even in the early efforts of the League of Nations to collect and publish arms trade statistics, the resulting statistical yearbooks were of some value. Naoum Sloutsky has written his recollections of this period, when he was the chief editor from 1920 to 1939: "It was almost impossible to find two nations whose customs systems were sufficiently similar to permit the use of any uniform control procedure. Hence, the *Statistical Year-Book* on arms and ammunition trade could not be regarded as a valid control instrument. It did, however, as we pointed out, provide valuable help in detecting the armaments race and describing its pace."[14] Those of us laboring to discern the patterns and purposes of the arms trade some fifty years later are also left to muddle through with "dirty" data; perhaps the best phrase to use in suggesting an antidote to the data problem is "multiple streams of evidence."[15] Some of the problems are problems of reliability, others are of validity. To the extent that analysis and theory building can be improved, it will rely on recognizing that the data that are available must be used so that they fit the class of propositions being explored. It is this problem that is the major focus of this chapter.

## The Data

### A Short History

Data on the arms trade have a relatively short history, especially compared with other, less controversial phenomena. Historians of international affairs researching the past 150 years can find foreign trade statistics, detailed records of diplomatic exchanges, lists of combat fatalities, and a host of other data needed to tell the story of the evolving international system. Arms trade data, however, have their origins as a formal data collection in 1924 with the publication of the first edition of the *Statistical Year-Book of the Trade in Arms and Ammunition*. Unlike other data compilations, the yearbook was not created as a result of functional systemic requirements. Foreign trade and international financial dealings were impossible without data, the generation and publication of which was in the interest of all parties; not so with armaments. The effort to produce arms trade data was uniquely related to the aftermath of World War I and the perceived role of the arms trade in causing that and other armed conflicts.

Sloutsky's account is a truly fascinating eyewitness account of this effort. A reading of his memoir surfaces all of the measurement problems that still plague current analysts. For example, what is meant by the term *arms?* Trade statistics of that era did not distinguish between civilian and military aircraft. Small arms were much more critical to conventional war than is now the case. Major problems existed in distinguishing weapons for war and weapons for sport, since the same weapons performed both functions. The recipient countries were very unsupportive when it came to providing data on the arms they actually received, especially since arms-producing countries were not required to provide data on arms production, even to their own governments. It should also be noted that the impetus for the data collection effort was the creation of a *new* international system in which the arms trade would be controlled, not for the purpose of ensuring a smoothly operating (and presumably brisk) trade in armaments. This was the merchants-of-death era, and the data problems reflected this perception. And yet, data were collected, and in Sloutsky's view, international public opinion and eventually the international system were changed as a result of this data collection and publication.

This very brief review of the original data-collection effort demonstrates the continuity of dilemmas in analyzing this important international phenomenon. However, it would be misleading to conclude that today we do not have available data that are infinitely more valid and reliable for use in assessing the role of arms transfers in world politics. Intelligence-collection capabilities are significantly improved, and the publication of the results (albeit for national purposes on occasion) has provided us with a great deal more information on military affairs in general. The information revolution has seen organizations such as the International Institute for Strategic Studies (IISS) and its publication *The Military Balance* accepted as producing a very close approximation of reality when it comes to military equipment inventories. SIPRI has published its arms trade registers continually since 1969 and has also closely approximated the arms trade reality. I now turn to a brief review of these data collections.

## The Sources of Data

Brzoska put forth three types of sources of data on the arms trade.[16] The first source of information is the arms industry. The most respected public source of information on military equipment is Jane's Publishing Company, which now has a virtual monopoly in this field. Jane's annual books on all types of equipment, by country, would not be possible without maximum information from industry. It is obviously in the interest of industry to get its products into the public arena; but although it is true that information about weapons system *characteristics* is provided by industry, very little is provided about actual production rates, and almost nothing about exports. The glossy photographs in defense-oriented journals let us know when there is a new product, but rarely to whom it is being sold. One reason industry members do not provide more information on their export activity is the traditional one of proprietary information and the fear of competitors. But there is another reason that is related more to the specific history of controlling these commodities: despite a general lack of success in halting or reducing arms trade in the 1930s, the effort did succeed in having national governments gain control over the export of arms from their own private industries. As a result, industries in general feel that since they have reported to their governments, they have met their responsibilities with regard to publishing export data.

Recipient governments are a second source of information. For example, the Soviet Union published no data on its arms exports, yet a great deal of data exist on what the USSR exported to whom. These data are almost exclusively gained by access to sources within the recipient governments, however obtained. This is particularly true when recipients are acquiring arms for public purposes such as prestige or deterrence.

Intelligence agencies are a third source of information. For example, the annual series on *World Military Expenditures and Arms Transfers,* published by the U.S. Arms Control and Disarmament Agency, is an unclassified version of statistics generated by the CIA. The foreword to a recent edition puts forth an altruistic motive for the publishing of these data: "This series has helped analysts and scholars for over two decades overcome the general scarcity and uneven quality of publicly available data on world military spending and arms transfers. . . . [It provides] important information concerning the military priorities and participation in the international arms trade of all nations."[17] But government intelligence agencies have other motives as well. For example, for the past few years, the CIA has been revising Soviet arms transfer data upwards and going back in time to change the records of those data also.[18] The result was a public source that provided evidence that the primary adversary had not slacked off in its arms exports, a source that was not unimportant in policy debates. Without Western intelligence agencies publishing even limited and selective data about the arms exports of Eastern bloc countries, a major gap in overall knowledge of the arms trade would have existed.

A fourth source is the supplying government. As Sloutsky and his colleagues in the 1930s discovered, national foreign trade statistics are designed to conceal, not reveal, data on arms exports. If anything, the problem is worse today.[19] But as Kolodziej's definitive study of French arms transfers demonstrates, diligent research of indigenous government sources can reveal valuable information.[20] The intricate notification and control system of the United States provides numerous opportunities to generate data, particularly from congressional sources. Since 1982, the Congressional Reference Service (CRS) has been distilling and integrating U.S. government data and publishing annual statistical reports on arms exports to the Third World, the most recent being published in July 1991.[21] The U.S. Defense Security Assistance Agency (DSAA), the bureaucracy responsible for administering U.S. arms exports, maintains an intricate data base and provides it to the public in its annual publication *Foreign Military Sales, Foreign Military Construction Sales and Military Assistance Facts*. Annual and bilateral data, critical to trend analysis, are available in current dollars, but actual numbers and types of equipment are not provided except in very aggregated form. In general, the availability of government information on arms transfers is a function of the openness of the political system and the size of the export effort, which explains the preponderance of U.S. data in the study of the arms trade.

A final source of information on the arms trade is the journals and other publications specializing in military affairs. A surprising amount of information is available on the basic unit of analysis, the arms deal. For example, I would estimate that if *Jane's Defence Weekly* was coded for individual arms transfer events, more than 50 percent of the world's arms trade would be captured in the resulting data base.[22] Other important journals and newsletters include *Defense News,* AAS *Milavnews Newsletter, Afrique Defense, Asia-Pacific Defence Reporter,* and *Defense & Foreign Affairs.* SIPRI lists over 60 periodicals and newspapers it uses to compile its arms trade registers, which are published annually in its *Yearbook on Armaments and Disarmament.*

So there is a wealth of information generated by a combination of professional investigative reporting and leaks (intended and unintended) of information from governments and industries. But the resulting data are skewed in some definitive ways. For example, data on actual delivery dates are always sketchy at best. Many times, the arrival of equipment does not enter the public "data base" until it shows up in the IISS publication *The Military Balance*. A good example was the transfer (illegal, it turns out) of eighty-seven Hughes 500D helicopters from the United States to North Korea. Any type of trade is prohibited between these two countries, let alone that involving arms. A listing in *The Military Balance* was the first concrete public evidence that these civilian helicopters had arrived and, as it turns out, were armed.[23] As previously mentioned, open-source data bases were always weak on Eastern bloc exports, except where the recipient desired it to be known. A classic case of this phenomenon is the past few years in Nicaragua. Hardly any major imports went unnoticed, since reporters were allowed (tolerated?) to observe the docks of Corinto and El Bluff, where the latest shipments could be recorded as they were unloaded.[24] U.S. intelligence provided more comprehensive data, for obvious reasons,[25] but the basic thrust of the arms transfer story was publicly known.

Another type of data generally missing in public sources is the activity that occurs early in a deal. These negotiations are usually very sensitive; rather, what we see is the deal surfacing at the formal negotiations stage, especially when two or more companies/countries are competing for an order. A final missing data point is the actual dollar value of the deal. Institutions such as SIPRI and Jane's have developed methods and techniques for assigning values, but even intelligence agencies have difficulties in uncovering actual contract figures on many deals.

To summarize briefly the sources of arms transfer data, it can be said that a great deal of information is available, but as in any intellectual enterprise, the data generated are a function of the scholars and analysts working in the field. As a rule, valid and reliable data will only be generated given a demand. To the extent that the analysis of the arms trade only requires a certain type of data, that will be the norm. We see this most clearly in the dominance of dollar values as a standard measure of arms transfer behavior, a measure to which I now turn.

## Arms Transfer Data Sets

The basic data sets available for use are produced by SIPRI, ACDA, DSAA, and the CRS. They have dollar values as their basic unit of measurement. Since the inception of these data sets, they have been exhaustively evaluated and compared.[26] They will be briefly summarized here so that the reader is aware of the basic approach each takes in capturing the essence of the international arms trade.

The annual ACDA publication *World Military Expenditures and Arms Transfers* (*WMEAT*) records the dollar value of arms exports and imports deliveries for every country in the world, going back to the last ten years and adjusting the data in each annual edition using a new constant dollar baseline (figure 2–1). *WMEAT* also aggregates supplier-recipient data on a five-year basis. For example, it would show the amount of arms delivered from the USSR to Angola for the most recent five-year period, but not the annual dollar amount. Only the top ten suppliers are listed in this table, so much of the so-called South-South arms trade among the new suppliers is not available. ACDA makes it clear that it is counting only deliveries, and it cautions users regarding various valuation techniques used. For example, the most recent edition talks of the reevaluation of Soviet data, and also the different method used to calculate dollar values for Soviet bloc exports. In the main, the dollar values in *WMEAT* represent contract values.

In an earlier edition of the publication, ACDA defended the use of dollar values in responding to the criticism that not enough information was being made available, particularly data on the actual equipment being delivered:

> A comparison of the arms trade of various countries by numbers of physical units can be useful for obtaining a rough measure of relative military

## Value of Arms Transfers and Total Imports and Exports, 1978-1988
### By Region, Organization, and Country

| YEAR | ARMS IMPORTS [a] Million dollars | | ARMS EXPORTS [a] Million dollars | | TOTAL IMPORTS [b] Million dollars | | TOTAL EXPORTS [b] Million dollars | | ARMS [c] IMPORTS TOTAL IMPORTS | ARMS [c] EXPORTS TOTAL EXPORTS |
|---|---|---|---|---|---|---|---|---|---|---|
| | Current | Constant 1988 | Current | Constant 1988 | Current | Constant 1988 | Current | Constant 1988 | % | % |
| Ethiopia | | | | | | | | | c | |
| 1978 | 1500 | 2519 | 0 | 0 | 455 | 764 | 306 | 514 | 329.7 | 0 |
| 1979 | 330 | 509 | 0 | 0 | 567 | 875 | 418 | 645 | 58.2 | 0 |
| 1980 | 775 | 1096 | 0 | 0 | 722 | 1021 | 425 | 601 | 107.3 | 0 |
| 1981 | 430 | 555 | 0 | 0 | 739 | 954 | 389 | 502 | 58.2 | 0 |
| 1982 | 575 | 697 | 0 | 0 | 786 | 953 | 404 | 490 | 73.2 | 0 |
| 1983 | 975 | 1138 | 0 | 0 | 876 | 1023 | 403 | 471 | 111.3 | 0 |
| 1984 | 1200 | 1351 | 0 | 0 | 928 | 1045 | 417 | 469 | 129.3 | 0 |
| 1985 | 775 | 847 | 0 | 0 | 993 | 1085 | 333 | 364 | 78.0 | 0 |
| 1986 | 330 | 352 | 0 | 0 | 1102 | 1174 | 455 | 485 | 29.9 | 0 |
| 1987 | 1000 | 1033 | 0 | 0 | 900 | 930 | 390 | 403 | 111.1 | 0 |
| 1988 | 725 | 725 | 0 | 0 | 900 | 900 | 390 | 390 | 80.6 | 0 |

**Figure 2-1. Excerpt from *World Military Expenditures and Arms Transfers 1989***

Source: U.S. Arms Control and Disarmament Agency, *World Military Expenditures and Arms Transfers 1989.*

capabilities. However, the variations in characteristics of systems produced by different countries may limit severely the validity of such comparisons. The variation in the military worth of particular types of systems in various environments further complicates such comparisons.[27]

This does not mean that data on types of equipment are not made available. For example recent editions of *WMEAT* contain data on the number of tanks, combat aircraft, and so forth (ten types of armaments in all) delivered by each major supplier to the developing world and to each major region (e.g., Africa) for the most recent five-year period. But these data are at a level of aggregation that makes the trend analysis that would illuminate systemic shifts much more difficult.

Since 1969, SIPRI has been publishing its annual *Yearbook on Armaments and Disarmament,* which contains dollar values for arms trade (figure 2–2). These dollar values differ in two major ways from the ACDA data. First, SIPRI only tracks and values major weapons systems, whereas ACDA tracks not only

**Values of imports of major weapons by the Third World: by region, 1970–89**[a]

Figures are SIPRI trend indicator values, as expressed in US $m., at constant (1985) prices. A = yearly figures, B = five-year moving averages.[b]

| Region[c] | | 1970 | 1971 | 1972 | 1973 | 1974 | 1975 | 1976 | 1977 |
|---|---|---|---|---|---|---|---|---|---|
| South Asia | A | 857 | 1 274 | 1 800 | 1 049 | 936 | 584 | 1 066 | 1 932 |
| | B | 1 135 | 1 181 | 1 183 | 1 129 | 1 087 | 1 113 | 1 278 | 1 376 |
| Far East | A | 2 299 | 3 582 | 6 962 | 1 815 | 1 920 | 1 595 | 1 490 | 1 983 |
| | B | 3 697 | 3 329 | 3 316 | 3 175 | 2 757 | 1 761 | 2 154 | 2 958 |
| Middle East | A | 5 242 | 6 092 | 5 842 | 10 472 | 6 999 | 7 014 | 7 076 | 9 816 |
| | B | 4 813 | 6 179 | 6 930 | 7 284 | 7 481 | 8 276 | 7 716 | 7 560 |
| North Africa | A | 185 | 224 | 373 | 340 | 591 | 2 343 | 2 282 | 2 619 |
| | B | 258 | 293 | 342 | 774 | 1 186 | 1 635 | 2 354 | 3 386 |
| South America | A | 285 | 786 | 1 093 | 2 354 | 1 338 | 1 600 | 1 922 | 2 836 |
| | B | 628 | 1 033 | 1 171 | 1 434 | 1 661 | 2 010 | 2 006 | 2 066 |
| Sub-Saharan Africa | A | 389 | 441 | 266 | 466 | 869 | 645 | 1 044 | 2 562 |
| | B | 278 | 339 | 486 | 537 | 658 | 1 117 | 1 528 | 1 536 |
| Central America | A | 185 | 135 | 261 | 309 | 299 | 204 | 234 | 557 |
| | B | 140 | 191 | 238 | 242 | 261 | 321 | 312 | 312 |
| South Africa | A | 275 | 104 | 292 | 459 | 533 | 232 | 371 | 171 |
| | B | 181 | 240 | 333 | 324 | 378 | 353 | 330 | 244 |
| Total[d] | A | 9 717 | 12 639 | 16 890 | 17 263 | 13 486 | 14 217 | 15 485 | 2 477 |
| | B | 11 130 | 12 784 | 13 999 | 14 899 | 15 468 | 16 586 | 17 679 | 9 436 |

[a] The values include licensed production of major weapons in Third World countries (see appendix 7C). For the values for the period 1951–69, see Brzoska, M. and Ohlson, T., SIPRI, *Arms Transfers to the Third World, 1971–85* (Oxford University Press: Oxford, 1987).

**Figure 2–2. Excerpt from *SIPRI Yearbook 1990.***

smaller items but also ammunition, support equipment, and anything else used for war. Secondly, SIPRI's "dollars" represent more than contract value. For example, they contain a military use component, as well as a modernity component reflecting different values for refurbished and secondhand equipment.[28]

The DSAA's data set, *Foreign Military Assistance and Sales Facts,* is easy to understand. For each country that is a recipient of the many forms of U.S. security assistance (commercial sales, FMS credits, and so on), the dollar amount in current dollars is listed (see figure 2-3). Data are available going back to the 1950s. Because these data are in current dollars, some types of comparisons are not possible without conversion by the analyst. Annual data on the types and numbers of equipment transferred on an annual basis are not available, because they are classified. DSAA does release this type of information on an aggregated basis.

The most recent entry into the data set field is the CRS. Its annual publication lists dollar values (some current and some constant) on the arms trade between major suppliers and recipient regions of the Third World. These dollars are the same as used in the DSAA and ACDA data (that is, contract values as generated by the intelligence community; see figure 2-4). The unique feature of this data set is its integration of U.S. and foreign arms suppliers into one data base on an annual basis.

For the scholar and analyst who wishes to get beyond dollar amounts to the actual deal itself, only SIPRI has consistently provided this type of data. Its annual arms trade registers provide information as to the date of agreement and delivery, number and type of weapon systems involved, and payment and production arrangements, if known. A sample entry is shown in Figure 2-5.

As mentioned earlier, the collection problems inherent in this field make these entries vary significantly in terms of validity and reliability. Another option is to use the annual editions of *The Military Balance,* in essence comparing one year's inventory with the next and assuming new items were delivered during the year. This technique would have to be controlled for the possibility of indigenous production, accidents, retirements, and other causes of inventory changes. However, it is one of the few methods available to track inventory changes attributable to the arms trade. This approach, of course, would not capture a vital part of the arms trade (particularly as it applies to world politics), which is agreements reached during a particular year. *WMEAT* goes out of its way to tell the user that deliveries are a more useful indicator, although recent editions have begun to list agreement data at an aggregated level that allows some systemic analysis.

## Measurement Problems

How are we to describe, explain, and forecast international arms transfer systems and the national policy response? How are we to test propositions that

(DOLLARS IN THOUSANDS)

| | FY 1950 FY 1979 | FY 1980 | FY 1981 | FY 1982 | FY 1983 | FY 1984 |
|---|---|---|---|---|---|---|
| WORLDWIDE | 71,964,573 | 12,431,442 | 6,409,377 | 16,823,868 | 14,552,503 | 13,350,002 |
| EAST ASIA AND PACIFIC | 9,092,530 | 1,962,377 | 1,764,521 | 4,783,838 | 2,096,844 | 1,603,350 |
| AUSTRALIA | 2,282,477 | 395,950 | 408,243 | 2,649,313 | 152,379 | 391,070 |
| BRUNEI | 10 | | 149 | 106 | | |
| BURMA | 4,365 | 681 | 577 | 984 | 182 | 1,349 |
| CHINA | | | | | | 629 |
| FIJI | 87 | 1 | | 4 | | 1 |
| INDOCHINA | 8,542 | | 1,439 | | | |
| INDONESIA | 195,271 | 10,663 | 39,126 | 49,155 | 32,542 | 10,380 |
| JAPAN | 1,307,665 | 417,993 | 553,904 | 438,926 | 280,363 | 197,224 |
| KOREA (SEOUL) | 1,929,724 | 315,426 | 251,620 | 889,614 | 333,889 | 158,042 |
| MALAYSIA | 98,329 | 23,114 | 37,136 | 1,493 | 3,936 | 2,206 |
| NEW ZEALAND | 142,339 | 14,425 | 12,096 | 16,576 | 10,321 | 12,506 |
| PAPUA NEW GUINEA | | | | | | 143 |
| REP OF PHILIPPINES | 185,218 | 10,543 | 6,092 | 15,179 | 17,949 | 11,584 |
| SINGAPORE | 166,757 | 110,810 | 34,581 | 27,646 | 380,047 | 11,376 |
| TAIWAN | 2,015,255 | 455,449 | 286,383 | 522,776 | 695,502 | 704,461 |
| THAILAND | 762,324 | 216,324 | 133,175 | 172,066 | 189,734 | 102,380 |
| VIETNAM | 1,167 | | | | | |
| NEAR EAST AND SOUTH ASIA | 35,938,647 | 5,883,041 | 1,222,640 | 2,122,686 | 5,038,626 | 4,162,629 |
| ALGERIA | 174 | | 55 | | | 79 |
| BAHRAIN | | 4,863 | | 4,492 | 5,388 | 162 |
| BANGLADESH | | | | | 28 | 386 |
| EGYPT | 634,019 | 1,868,832 | 288,794 | 1,556,211 | 673,735 | 841,441 |
| INDIA | 76,208 | 1,019 | 184 | 828 | 10 | 528 |
| IRAN | 11,169,830 | | | | | |
| IRAQ | 13,152 | | | | | |
| ISRAEL | 8,114,907 | 513,513 | 116,781 | 553,538 | 2,140,879 | 103,257 |
| JORDAN | 892,788 | 288,441 | 276,752 | 94,625 | 44,363 | 23,640 |
| KUWAIT | 648,515 | 120,474 | 42,750 | 111,645 | 145,147 | 101,696 |
| LEBANON | 62,578 | 30,002 | 44,893 | 7,479 | 323,947 | 100,554 |
| LIBYA | 28,249 | | | | | |
| MOROCCO | 394,820 | 251,483 | 25,491 | 10,235 | 63,122 | 32,195 |
| NEPAL | 73 | | | | | |

**Figure 2–3. Excerpt from DSAA Publication** *Foreign Military Sales, Foreign Military Construction Sales and Military Assistance Facts,* **September 1989.**

Source: U.S. Defense Security Assistance Agency, *Foreign Military Sales, Foreign Military Construction Sales and Military Assistance Facts* (Washington: U.S. Department of Defense, 1989.

## ARMS TRANSFER AGREEMENTS WITH THE THIRD WORLD, BY SUPPLIER
(In millions of constant 1989 U.S. dollars)

| | 1982 | 1983 | 1984 | 1985 | 1986 | 1987 | 1988 | 1989 |
|---|---|---|---|---|---|---|---|---|
| Non-Communist | | | | | | | | |
| Of which: | | | | | | | | |
| United States | 12,950 | 9,825 | 7,621 | 5,508 | 4,273 | 5,760 | 9,301 | 7,718 |
| France | 8,301 | 2,027 | 7,649 | 1,713 | 1,404 | 3,293 | 3,179 | 300 |
| United Kingdom | 1,725 | 808 | 750 | 9,925 | 895 | 547 | 5,178 | 3,200 |
| West Germany | 1,175 | 639 | 597 | 193 | 520 | 847 | 83 | 1,290 |
| Italy | 1,463 | 1,315 | 808 | 1,463 | 542 | 118 | 166 | 240 |
| All Other | 4,488 | 7,467 | 3,971 | 3,993 | 5,229 | 2,510 | 2,392 | 2,650 |
| Total non-Communist | 30,103 | 22,081 | 21,396 | 22,795 | 12,862 | 13,075 | 20,299 | 15,398 |
| Communist | | | | | | | | |
| Of which: | | | | | | | | |
| U.S.S.R. | 26,153 | 8,118 | 24,868 | 18,705 | 17,997 | 23,187 | 14,654 | 11,230 |
| China | 1,975 | 1,001 | 398 | 1,599 | 1,979 | 4,998 | 2,392 | 1,120 |
| All Other | 3,200 | 3,329 | 890 | 4,855 | 5,240 | 2,381 | 2,454 | 1,530 |
| Total Communist | 31,329 | 12,449 | 26,157 | 25,159 | 25,216 | 30,566 | 19,501 | 13,880 |
| GRAND TOTAL | 61,431 | 34,530 | 47,553 | 47,954 | 38,078 | 43,641 | 39,800 | 29,278 |

Figure 2-4. Excerpts from CRS Publication *Trends in Conventional Arms Transfers to the Third World by Major Supplier, 1982-1989,* August 1990.

Source: U.S. Congressional Research Service, *Trends in Conventional Arms Transfers to the Third World by Major Supplier, 1982-1989* (Washington: Library of Congress, August 1990.

| Recipient | Supplier | No. ordered | Weapon designation | Weapon description | Year of order | Year(s) of deliveries | No. delivered | Comments |
|---|---|---|---|---|---|---|---|---|
| Lesotho | Spain | 2 | C-212A Aviocar | Trpt aircraft | (1988) | 1989 | 2 | |
| | USA | 1 | Model 182 | Lightplane | (1989) | 1989 | 1 | but never delivered |
| Libya | France | 2 | Mirage F-1A | Fighter/grd attack | 1986 | 1989 | 1 | |
| | | 2 | Mirage-5 | Fighter | 1986 | | | |
| | USSR | 1 | Il-76 Candid | Trpt aircraft | (1988) | 1989 | 1 | |
| | | (15) | Su-24 Fencer | Fighter/bomber | (1988) | 1989 | (12) | |
| | | .. | SA-5 SAMS | Mobile SAM system | 1989 | | | Part of SA-5 air defence system |
| | | .. | Square Pair | Tracking radar | (1988) | | | |
| | | .. | AS-14 Kedge | ASM | 1989 | | | Arming Su-24 Fencers |
| | | .. | SA-5 Gammon | SAM | (1988) | | | |
| | Yugoslavia | 4 | Koncar Class | FAC | 1985 | | | |
| Malaysia | France | 1 | Falcon-900 | Trpt aircraft | 1988 | 1989 | 1 | For VIP use |
| | Italy | 4 | Skyguard | Air defence radar | (1987) | 1988–89 | (4) | Fire control for 1 bty of 35-mm anti-aircraft guns from Switzerland |
| | Netherlands | 1 | Flycatcher | Mobile radar | 1988 | 1989 | 1 | |
| | UK | 8 | Tornado IDS | MRCA | 1989 | | | Deal incl artillery, SAMs, radar and 1 submarine subject to final negotiation |
| | | 6 | Wasp | Helicopter | 1988 | 1989 | 6 | Second order |
| | | 30 | FH-70 155mm | Towed howitzer | 1988 | 1989 | 9 | |
| | | (24) | L119 105mm gun | Towed gun | 1988 | | | |

Figure 2-5. Sample Entry from Arms Trade Register, *SIPRI Yearbook 1990.*

Source: Stockholm International Peace Research Institute, *SIPRI Yearbook 1990.*

shed light on the effects of these transfers? Only through the use of valid and reliable data can we avoid the problem plaguing every intellectual endeavor of this type, namely, generating conclusions that are more a function of data than of the underlying realities.

## Reliability

Arms are not as benign as tractors and automobiles, and conclusions reached regarding the effects of trading these commodities are therefore subject to much more scrutiny and debate. One of the first targets for criticism of a research conclusion is the reliability of the data (that is, actual measurement errors caused by the measurement practices employed). One data point in error can bias an audience against all of the conclusions, regardless of the rigor.

The first issue of reliability concerns the counting rules, a problem that has plagued intelligence agencies, customs officials, reporters, and anyone else involved in monitoring the arms trade. Governmental trade statistics suffer from this problem. "One major problem with such data is the absence of standardization. Governmental bodies use their own definitions of weapons and in many countries, for example, Brazil or FR [West] Germany, various figures are available reflecting different definitions. Compilations of such figures are not reliable."[29] Counting weapons systems is particularly susceptible to reliability errors; for example, how are weapons in storage handled? Different agencies and organizations develop and utilize different counting rules. When the data produced are mixed, matched, and compared, the result can be unreliable data.

Dollar valuation is also ripe for reliability errors. Contracts are valuable information and are seen as primary evidence that an agreement has been made to transfer arms. Yet, these contracts are hard to come by. Various methodologies are developed to approximate the value of a deal involving twenty tanks going to country X based on the contract price of this type of tank in a previous deal with country Y. Many aspects of a specific deal may make such estimates erroneous. Additionally, until recently we had the major controversy over how to evaluate USSR and Warsaw Pact arms transfers, given the problem of trade in rubles. ACDA warned its readers in this regard: "Particular problems arise in estimating the expenditures of communist countries. . . . Data on Soviet military expenditures are based on CIA estimates of what it would cost in the United States in dollars to develop, procure, staff and operate a military force similar to that of the Soviet Union. Estimates of this type—that is, those based entirely on one country's price pattern—generally overstate the relative size of the second country's expenditures in intercountry comparisons."[30] Brzoska's detailed assessment of the *WMEAT* data contains three pages alone on their reliability; he concludes that "they are not Soviet prices at all, but U.S. prices. While the ACDA pricing method does in theory correctly reflect military-use value, it clearly overstates the economic costs to the USSR and to the arms recipients. It is not surprising that 'Soviet exports' and 'U.S. dollars costs' differ by a wide margin."[31] While the Soviet Union as an arms supplier has disappeared from the system, eliminating this pricing controversy, it remains true that valuating arms transfers remains a serious methodological problem.

Reliability of arms transfer data is not enhanced by the overall secrecy surrounding the trade of these sensitive commodities. At the international level, national security demands that each country be very careful regarding the release of arms transfer data. Because these commodities can have a major political impact, governments go to great lengths to conceal and distort information. This is especially true at the initial and terminal stages of a deal. This can be seen with the U.S. Congress, which, in its effort to play a more forceful role in this policy area, has continually demanded more information regarding who the executive branch is talking with about arms sales. These data exist in the form of price and availability data that are sent to prospective customer governments, in essence telling them what they will have to pay for the item desired and when they can have it.[32] But these data are not publicly available in the timely fashion needed to affect policy at the early stage of a deal. Debates over whether or not the United States is really controlling arms exports often comes down to Congress demanding to know if the government has ever turned any country down in its request for arms. The executive branch rarely releases information about specific cases because of the presumed diplomatic damage to the would-be recipients.[33] And the United States is considered the country most free with its arms transfer information.

Intranational secrecy stemming from bureaucratic politics also contributes to data problems, as the above example of price and availability data indicate. The bureaucratic competition, even in a fairly open system such as the United States, leads to the withholding and distortion of information, inevitably resulting in less reliable arms trade data. Counting rules not only vary from nation to nation and from institute to institute; they can vary widely within a government. Intelligence agencies, for example, establish elaborate rules regarding the reliability of evidence concerning arms deliveries. The dollar-valuation problems mentioned earlier are a product of different bureaucratic orientations, and the United States is hardly alone in this matter. The bureaucratic politics in Germany are such that only companies of a certain size must report export data, but most of the exports come from the very companies that do not have to report.[34] Much of the unreliability in national-level statistics on the arms trade, therefore, can be blamed on the nature of the national security and foreign policy processes within the country and not necessarily on the rational-actor model, which posits deceit, deception, and politically motivated data alterations by the state itself.[35]

It must also be noted that data and values can change as a result of the research effort and emphasis itself. As already mentioned, data on Soviet arms transfers for 1979–88 in *WMEAT* were altered for this reason: "The estimates of Soviet arms transfers in value terms used in this and the previous edition have been revised upward substantially from earlier editions. The revisions generally do not affect underlying estimates of the number, type, or value of major military equipment deliveries, but rather the estimated dollar value of supporting material deliveries, particularly those to countries engaged in ongoing hostilities."[36] In effect, ACDA looked harder at the raw data available and found more arms transfers. The definition of arms transfers did not change; only the amounts discovered with increased effort. In effect, this means that the reliability of the *WMEAT* data will vary considerably from country to country as a function not only of reality but of the effort expended to uncover evidence.

Kolodziej's classic case study of the ACDA data in 1979 concluded that ACDA had seriously underestimated French arms transfers, based on his examination of primary data from French sources. By the time he published his major book on arms transfers eight years later, he had concluded that "the major finding of the article—that French estimates of the economic value of its arms deliveries were higher by a factor of two to three when compared to ACDA data—has gradually been narrowed in the 1980s as ACDA has raised its estimates of the value of French deliveries, bringing them into closer alignment with official French sources."[37] This may well be the result of increased collection and analytical effort corresponding to France's rise as a critical arms supplier.

A final reliability issue is that collectors of data do not have access to or use the same sources, resulting in serious confusion for those analyzing arms transfer trends. Kolodziej addresses this problem of reconciling sources at the end of his book, identifying three key problems. First, there are contradictions within the same source over time (or between different sources) concerning dates and quantities of deliveries. A second problem is the lack of information on weapons, for example, using a common name (such as "Sidewinder") but failing to differentiate among model numbers and their vastly different military capabilities and price. A third pattern of problems relates to the question of who ordered, who paid for, and who supplied what.[38] As Kolodziej suggests, these reliability problems are not insurmountable, given the appropriate collection effort. It must be concluded, however, that the problems remain as long as the paradigms and puzzles that encompass this field do not demand more than the data effort that currently exists. It would be fruitless to improve the data—for example, ensuring that the model numbers accompanied each case—if researchers did not feel an intellectual need for such precision. In essence, this is mainly a problem of validity and matching propositions with appropriate data, the next issue to be addressed in this chapter.

### Validity

The validity of data refers to the extent to which they represent the phenomenon or concept being studied. As previously stated in this chapter, the term *arms transfer* has many dimensions. Validity problems occur, for example, when the analyst is monitoring flows of military capability being transferred without including dual-use equipment (such as computers and electronic equipment) that adds to the military capability of the recipient. Validity problems can occur when the economic effects of the arms trade are evaluated using contract data that may not reflect monies actually paid by the recipient.

Ensuring the validity of data starts with defining what is meant by the term *arms*. It can mean the actual end item (for example, a tank), dual-use equipment that enhances the capability of the end item (wheeled flatbed trucks that can serve as tank transporters), arms production technology (a license to produce gun barrels in the recipient country), or support services (ammunition for the tank). This is not a reliability problem, since all of this information may be available and accurate, but not used on the grounds that it does not fit the definition of "arms" for a particular analyst, bureaucracy, or institution. This problem is as old as the arms transfer control issue itself, as the 1930s saw most states develop "munitions lists" of items to be controlled. And definitions of arms are not trivial in a policy sense; one has only to observe the hue and cry of industry when an item is added to or dropped from these lists.

Social scientists refer to *construct* or *face* validity as that property of data that allows us to be confident that they are truly measuring a dimension (for example, the military utility of an arms transfer), because we can correlate the data with some other measure. A typical method of ascertaining the level of construct validity is to use experts. For example, can we use dollar values as a surrogate measure of military capability? One way to determine their validity is to construct rankings of states on this dimension and ask experts if the ranking reflects reality as they see it with their own sets of data and analyses. Such an exercise would reveal serious questions as to the use of monetary values to test propositions involving military capability.

Historically, defense budgets have been used as a measure of a country's military capability. Richardson's landmark study has spawned several generations of analysts who use budget data to describe and explain arms races.[39] The basic rationale for using defense expenditure data to reflect military capability lies in the belief that weapons capability is related to cost. Even if cross-national budget data were available for weapons procurement alone (which it is not), however, serious validity problems ensue. Rattinger has observed that for Middle Eastern countries, "military spending is a rather meaningless indicator of capability because of the complex interplay of military aid, regular arms procurement, gifts, and nonmaterial forms of payment by political allegiance to arms donors and the like."[40] And there are other problems with using budget data: countries can misrepresent the data, overall budgets do not reveal allocation realities (e.g., personnel versus equipment), and there are the usual problems of inflation and currency transformations.

The dollar value of a deal may also be misleading when it is used in a comparative sense, because even accurate contract data cannot put a value on aspects of the deal that may be crucial to its military value. Saudi Arabia has acquired "arms" valued at roughly $31.6 billion[41] in the past twenty years from the United States, much of which is in the form of infrastructure, training, and logistics. Israel, on the other hand has a much lower total ($13.5 billion), but it would be an obvious mistake to conclude that the United States has been "tilting" toward Saudi Arabia, especially on the military dimension. Yet, the policy debates in the U.S. Congress during this period would appear to have been greatly influenced by this very basic and faulty assessment (that dollar values equate to military capability).

Modern fighter aircraft are considered a bellwether of arms trade relationships. They are expensive, visible, and get a great deal of attention in the policy-making process. Indeed, the U.S. Congress wrote restrictive legislation singling out this weapon system for special attention. But it is questionable whether tracking this system and its costs really gives insights into military capability and its effect, for example, on generating or exacerbating conflicts. Latin American countries are very fond of purchasing modern fighter aircraft that seldom are used in actual conflict, since they are either inoperative due to lack of spare parts or inapplicable to the low-intensity conflict that dominates that area. If anything, it may even be suggested that attention paid to these high-prestige, expensive systems in fact contributes to the control of conflict. The effect on economic development, however, may very well be different, and for these reasons costs are crucial. But it is clear that costs as an indicator of military capability can lead to misleading conclusions; many of the other dimensions listed above must be included.

What about the opposite approach, using various indices to measure the military capability of arms transfers? How valid, for example, is it to count numbers of tanks and other types of weapons systems exported? Even though major weapons can be counted easily, correlating inventories with capability may be less than totally valid because qualitative differences exist among individual weapons and among countries' capabilities to use the weapons. There are many examples of lighter-armed forces winning a battle. This inventory approach overlooks situational variables such as terrain, user operating proficiency, logistics, and the potential adversary's weapons and capabilities. But it is a clear improvement over dollar values, and I would suggest that the aforementioned panel of experts would be in more agreement with these rankings than those generated using monetary indicators.

When we observe the policy-making process in the major arms-supplying countries, we can see that strict inventories are not adequate in their eyes, either. Attention is indeed paid to specific capabilities that may be destabilizing, prestigious, and so forth; this is particularly true of the technological dimension. "Black boxes" on fighter aircraft (such as radars) are included or withheld in the name of politics or arms control. The shah of Iran was only interested in the latest military aircraft, the result being the sale of F-14s with Phoenix missiles at the same time they were entering the U.S. inventory. And yet he was denied a very sophisticated electronic-warfare aircraft, the Wild Weasel, when the U.S. military protested that it was too advanced to risk losing the technology to the

USSR should the shah depart the scene. Not surprisingly, therefore, indices have been developed over the years that attempt to integrate these factors into military capability indices.[42]

How have these approaches fared on the validity dimension? SIPRI has developed a dollar-valuation scheme that includes more than simple contract monetary values. For example, they use parametric estimation techniques in which an aircraft's speed and weight are regressed against cost. Also taken into account are spare parts accompanying the transfer, as well as the modernity of the weapon system.[43] It is difficult to assess the validity of this approach, since SIPRI does not publish the values generated for each weapons system, only the aggregate data in dollars. It is a useful indicator because it goes beyond mere monetary values, but it appears to fall short for use in military balance analysis.

What would it take to capture the essence of an arms transfer in terms of its military utility? Clearly, one would have to go beyond mere bean counting (as done in *The Military Balance,* for example) and the SIPRI modified-dollar approach. What was attempted by Laurance and Sherwin was the application of a method called *multiattribute utility* to the problem of developing valid indicators for military utility.[44] The method had several basic steps. First, military capability had to be defined in mission terms: military for what? Then one asks, what types of equipment contribute to the accomplishment of the mission? Given these systems, what characteristics are important to observe in a given weapons system type and model so that it can be compared to another? How can all of these data be aggregated to come up with a single value of military utility for a specific weapons system? What is the appropriate way to aggregate scores of weapons systems into a country score? Finally, how does one modify this capability score by the ability of the personnel in each country to operate the equipment?

The mission selected for analysis was sea denial, the ability of a country's navy to defend itself against attack from the sea. Submarines and missile patrol craft were the weapons systems whose characteristics were to be observed. Experts were used to value and aggregate values, resulting in country scores. A sample is shown in Table 2-1.

What can be concluded from this approach? First, it is a very tedious proc-

### Table 2-1
### Sample of Multiattribute Utility Scores

|  | Personnel Factor Weight |  | Personnel Score |  | Platform Weight |  | Platform Score |  | Capability Score |
|---|---|---|---|---|---|---|---|---|---|
| Egyptian OSA | (8.9 | × | 12.28) | + | (8.0 | × | 18.41) | = | 256.67 |
| Syrian OSA | (8.9 | × | 8.33) | + | (8.0 | × | 18.41) | = | 221.42 |
| Israeli SAAR | (8.9 | × | 20.00) | + | (8.0 | × | 15.49) | = | 301.92 |

Source: Sherwin and Laurance, "Arms Transfers and Military Capability," pp. 360-89.

40 • *The International Arms Trade*

ess involving experts who may be biased and not available in a real-time application of this method. However, the method appeared to generate a ranking that experts agreed reflected reality.[45] If generated on a regular basis, it would allow more valid conclusions to be reached regarding the effect that the transfer of specific weapons systems, in specific numbers to specific countries, would have on the relative military capability of countries and regions. To the extent that military balances are important in explaining and forecasting conflict, the approach would be useful.

It is clear that validity is a major challenge to research on the concept of arms transfers. The intricacy and rigor involved in calculating military utility for just one mission area makes it clear that something less valid will have to be employed. And it is not clear that these less valid measures will be inadequate to the task of determining the international systemic shifts and policy effects of arms transfers that are the subject of this book. Much will depend on how the data currently available are aggregated for analysis, which is the next topic in this chapter.

## Generating Cases and the Problems Associated with Aggregating Currently Available Data

Up to this point, the focus has been on the *variables* used in assessing the arms trade. An additional set of measurement problems can often occur as we select the *cases* to be used in the analysis. Conclusions can be significantly influenced by the specific cases from which generalizations are drawn. Observers of the arms trade are naturally drawn to those cases that are unusual or important in some way. Generalizing from a single case is both popular and dangerous; this is a particular problem when a government interested in justifying a particular arms transfer policy cites a case that "proves" the efficacy of the policy. Do arms transfers lead to conflict exacerbation or conflict resolution? There was no shortage of U.S. policymakers willing to cite the supply of Stinger missiles to the antigovernment *mujahadeen* in Afghanistan as evidence of the latter, in that Soviet troops were seen leaving the country as a result of the introduction of this weapon.[46] The contrary conclusion, that arms transfers can exacerbate conflict, is made by those who criticize the arming of the Nicaraguan *contras* or the acquisition of Soviet MI-24 armed helicopters by the Nicaraguan government. As George and others have pointed out, great care must be taken when using case studies to generate policy-relevant theory.[47]

This problem of case selection is no less when using the aggregated data sets discussed in this chapter, and may even be worse, since analysts are subject to the allure of numbers in columns appearing to be "true." Assuming that the analyst accepts the reliability and validity of these data, serious problems can arise when aggregating the data or artificially creating the cases to be used in testing propositions. For example, the common error of using current dollars for trend analysis has often produced misleading results. But the misled have only themselves to blame when arms transfers are depicted as "spiraling upward" on the basis of annual current dollar data that may reflect little more than the rate of inflation. But even when dollars are made constant, contrary conclusions can be reached using identical data but aggregating it for differing time periods. The most common problem is generating conclusions from a single year, and particularly the most recent year. Sudden spurts or drops in annual data can be put into perspective through a variety of techniques such as exponential smoothing and moving averages. This is particularly important when evaluating trends at the systemic level.

A final abuse of statistics occurs when data are selectively aggregated by a specific set of supplier and recipient countries. For example, most of the knowledge of the arms trade is really knowledge of the arms trade between major arms-producing states and the developing world. The conclusions reached differ significantly if trade among these industrialized states is the subject of the analysis.

The antidote for problems arising from the case selection and generation process is well-known. Observers and analysts of the arms trade are responsible for ensuring that conclusions and generalizations flow from the cases used, and that the data are aggregated in ways that do not produce misleading results. But even care in this facet of analysis cannot make up for misleading conclusions that stem from the improper and inappropriate generation of hypotheses and propositions, and the misapplication of data to these propositions. This chapter has presented the reality of data generation when studying the international arms trade. No attempt has been made to hide the very real problems of data reliability and validity that exist. However, the arms transfer phenomenon is central in the quest to describe and explain international security behavior, requiring those who study its effects and implications to use what is available while efforts continue to develop and refine more adequate data. In the final analysis, this involves the appropriate matching of data and propositions. This becomes very important, because what we know about arms transfers can be seriously affected by this match (or mismatch). As already shown, one of the areas where this is most problematic involves the use of dollars and budgetary data to operationalize arms transfers. The review in this chapter has made clear the availability of such data as well as their inherent attractiveness, especially compared to the tedious measurement practices required when using such techniques as multiattribute utility analysis. However, clearly misleading results can often be the end product.

## Methodological Challenges

The important concepts, propositions, and data that make up this field of study have been presented. What remains is to summarize the key measurement issues involved in arms transfer research before proceeding to the systemic level of analysis in the remainder of the book.

**Standard Measurement Challenges.** The first set of challenges stemming from the very nature of the international arms transfer system are those that transcend systemic changes. Arms are qualitatively different than commodities such as cars, wheat, or oil. It is no accident that ever since arms trade data were seriously collected in the 1930s, there has been no category called "military equipment" in any official trade statistics. This is not likely to change as long as both suppliers and recipients consider arms central to their national security. Secondly, as the international political system has begun to shift away from one dominated by East-West issues, regional conflict has been marked by the use of mid- to low-technology equipment. Armaments up to the level of small-caliber artillery, mortars, and rockets have always been difficult to track, but this reality becomes more critical to the generation of knowledge as their use becomes more prevalent in what is currently being termed *low-intensity conflict.* Many coups and small-scale invasions have been executed with imported arms totally invisible to analysts, even after the fact. Added to this problem is the increase in private and, in many cases, illegal arms exports beyond the control and scrutiny of governments, scholars, and policy analysts.[48] Lebanon's recent civil war involved roughly $1 billion per year in arms trade, most of it impossible to track.

Data on arms transfer agreements tend to be the most important indicator in terms of shifting relationships in the international system. But since these contracts involve industry and their orientation to proprietary information, hard data on the terms of the contract are difficult to obtain. A fourth recurring measurement issue resulting from the reality of the international system involves attaching a military value to any given transfer or set of transfers. Training, maintenance, logistics, and the ability of recipient personnel to operate the arms received become very important as the technology of newer weapons systems improves. There is little information available on these dimensions, let alone agreement among those studying the issue as to how to produce valid and reliable data.

**Challenges Created by New Developments in the International System.** A second basic set of challenges stems from those recent changes in the international system. Although chapter 6 will discuss these changes in depth, they will be introduced here to illustrate the need to develop measuring instruments constantly to cope with the propositions that make up the paradigm. It is important to point out that the measurement techniques and generated data discussed earlier in this chapter have their origin in the international arms trading system of the 1930s, which was dominated by state-centric trading in major end items (tanks, aircraft, ships, and the like.) Although state-centric suppliers still dominate the system, some of them are newcomers unrestricted by such 1930s developments as the use of end-user certificates restricting retransfer. What was previously an assumption (that all exports and retransfers were known to and controlled by governments) now has to be determined empirically. Additionally, suppliers no longer are one state. Multinational production is on the increase and muddies the waters, especially when the politics of the arms trade are at issue.

A second development involves the commodities being exported. As military capability becomes more a function of adding an upgraded component, the inherent size and source (smaller companies than the giants used to manufacture fighter aircraft) make this trade harder to track. This is one of the major revelations of the recent transfers from Germany to Iraq: most of the controversial trade was in fact legal but had escaped notice due to either its dual-use nature or its small size.

**The Special Challenge Presented by Offsets.** A third change with measurement implications involves the mode of payment. The most important development is the increasing use of offsets as an integral part of an arms transfer arrangement. "Offsets in arms trade are arrangements which use some method of reducing the amount of currency needed to buy a military item or some means of creating revenue to help pay for it."[49] Offsets can take several forms. The supplier may coproduce, subcontract, or license all or part of the weapons system in the recipient country; this means jobs in the recipient country that in former times would have stayed in the supplier country. Increasingly, offsets involve the transfer of technology to aid in the industrial development of the recipient country. Finally, they can involve extensive use of an instrument known as *countertrade,* in which the supplier agrees to market products of the recipient country in exchange for the purchase of military equipment.

This development can create difficulties for those observing and measuring the arms transfer phenomenon. First, governments can and do stay out of the offset arrangements, leaving industry to make the deals with a resulting diminution of public information. Secondly, some types of offset arrangements are inherently difficult to assess, even if the details are made known. How is technology transfer to be evaluated? Third, if the political impact of arms transfers are being assessed, the era of offsets requires considerably more information about the respective national military-industrial complexes.

These measurement difficulties result in significant implications for generating knowledge about arms transfers and their implications. For example, the traditional approach of assigning values to arms sales based on the announced contract is becoming less useful. The U.S. government may publish the value of 160 F-16 aircraft sold to Turkey, but without offset data from the Turks and General Dynamics (the makers of the aircraft), the economic magnitude of the transfer is seriously overstated. The most significant implication is that analysts must now look beyond the military capabilities of weapons systems in explaining and forecasting patterns of arms transfers. In the case of F-16s to Turkey, all of the aircraft competing for this contract were acceptable to Turkey, whose air force was very outdated. What surfaced as a nonmilitary—and critical—factor in the decision was the need for industrial development. General Dynamics agreed to help develop a Turkish aircraft industry, where none existed, in exchange for the contract. The Turkish desire for industrial and economic development, not the relative military capabilities of the competing aircraft, was the crucial factor in this case.

On the supply side of the offset equation, new issues also arise. The U.S. Congress is increasingly concerned with this issue, mainly because of the loss of U.S. business and jobs that is inherent in offset arrangements. This has prompted a major push for data on offset arrangements, data that have some serious flaws. Since 1985, annual reports on the magnitude of offsets have been issued by the U.S. government in an attempt to get a handle on their impact.[50] The data have fallen short of expectations for several reasons. First, they are generated based on a questionnaire distributed to and voluntarily answered by the industries involved. Industries ensured that the data did not include those details that could cut into their competitive position, the very details analysts need to assess properly the impact of offsets. One of the methods used was to aggregate the data and provide nothing by company or specific deal, precluding the dynamic analysis needed to test adequately the propositions that lead to the development of knowledge. Predictably, the Congress has begun to call for more government participation in offsets so that this information can be more available for congressional scrutiny and, presumably, more informed policy-making.

Does the generation and measurement of reliable and valid data matter? In August 1991, the Congressional Research Service released its annual report, consisting primarily of data on arms exports to the developing world, with very little analysis and commentary.[51] (See the section on "Arms Transfer Data Sets" earlier in this chapter for a description of this data set.) The picture that the data painted was one of a resurgent United States that had assumed the role of leading arms supplier. But unlike previous cold war years, when such an empirical reality would have gone unnoticed or perhaps been applauded, it did not fit with the larger U.S. policy perspective. The published data mattered in 1991 because the United States was involved in several arms trade control efforts (including the Iraq embargo, the Middle East peace process, and arms trade control talks in Paris with the other permanent members of the UN Security Council). Within two days of the CRS report, the State Department issued a statement rejecting the report as misleading because it was founded on dollar-value comparisons: "In dollar terms, the largest share of U.S. military exports are not for weapons but for construction, spare parts and support."[52] The State Department also noted that the U.S. dollar figures are higher than other countries because of more expensive U.S. technology and the fact that other countries' figures are understated as a result of subsidization of costs and barter. In short, the State Department cited most of the issues raised in this chapter about arms trade data.

# 3
# International Arms Trade as Systems and Regimes

## The Utility of the Systemic Approach

In chapters 1 and 2, the paradigm used in the study of the international arms trade was described in terms of the concepts, questions, puzzles, hypotheses, and data that have been used to generate knowledge about this issue area. In addition, it has been made clear that while this effort to generate knowledge can be made at multiple levels of analysis, little has been done at the level of the international system. As indicated in the preface, such an effort is the major objective of this book.

As the first book in this series demonstrated, a healthy discussion among scholars has been ongoing for some time regarding how the level of international system can be used to answer the critical question of change. Rosenau calls on "students of world politics to offer new understandings that are consistent with the emergent patterns and that thus shape how the world's politicians, journalists, and publics perceive the processes in which they participate."[1] In keeping with this challenge, it is the purpose of this chapter to explain how the systemic level of analysis will be used to delineate the several systems and regimes that have existed since the interwar period for one particular issue area, the international arms trade, emphasizing the transformation of systems and the effects on the actors in the system.

What is the payoff from such knowledge of arms transfer systems? This can best be seen by referring briefly to some examples of the puzzles that have emerged from the real world of policy actions and debates, and by seeing how knowledge at the systemic level has helped solve these puzzles and guide the making of policy. In the 1970s, the People's Republic of China began to break out of a long period of isolation, one that had produced a great deal of stagnation in the modernization of conventional military forces. There was a public confirmation of this reality as China began to search for weapons systems and military technology to acquire and develop for indigenous use. From which countries would it acquire such weapons? How would the transfers occur? Would knowing about the international arms transfer system of that period, the nature of the units and the environment, help in explaining its behavior? Much of the answer to these questions was found in knowledge of the international arms trade system extant at that time. For example, of the leading arms suppliers, only France and the United Kingdom had maintained any significant political contact with China. As such, they were rewarded with most of the arms transfer contracts signed in that period.

In the 1980s, both Brazil and Israel emerged as second-tier suppliers of military equipment. Both had developed indigenous products and were exporting them to a variety of clients. But there were very clear differences in the pattern of recipients. Brazil seemed free to supply its major defense systems to a wider variety of client, including both sides of the Iran-Iraq war. Israel, on the other hand, rarely exported its big-ticket items, such as the Kfir fighter and Merkava tank. One of the answers to this puzzle may lie in one of the rules of the system of that period, namely, the practice of the end-user certificate (which stipulates that arms exported to country A cannot be exported to country B without country A's permission). Israel's products, while extremely advanced and capable, were hampered when it came to exports, because many of them had U.S. components and their export was restricted by U.S. end-use restrictions. Brazil had no such restrictions, especially for such mid-level technology products as its line of wheeled armored personnel carriers that was the mainstay of its export activity, because they were manufactured with indigenous materials and technology.

Since the 1970s, South Korea has been constantly seeking to upgrade its fighter aircraft. In the early 1980s, it purchased F-16s from the United States as an off-the-shelf end item. But in the early 1990s, when it agreed to acquire the more modern F/A-18 fighter from the United States, it insisted upon and received a significant amount of technology and indigenous coproduction as part of the deal.[2] How is this change to be explained? Within South Korea itself, little had changed in the ten years between deals. For example, despite progress in other industries, its aircraft industry remained at the infancy stage. The answer lies with a systemic shift in the mode of transfer, from the acquisition of end items to offset arrangements as a route to diversifying dependence.

As a final example of how the systemic level can help in solving the puzzles of international arms trade, consider the plight of Iraq, which opened the war against Iran with equipment, spare parts, and ammunition supplied mainly by the USSR. When faced with a Soviet embargo, Iraq turned to France, China, and other second-tier suppliers. This response was significantly shaped by an international arms trade system governed by a structural hierarchy, rules regarding the retransfer of arms, and the control of conflict by the two major suppliers, the United States and the USSR. This case sparked a major debate regarding systemic transformation, particularly regarding the capabilities of the superpowers to control the effects of arms transfers. In the beginning phases of this war, Iraq was limited in where to obtain spare parts for Soviet equipment. That is less true today and, in effect, equates to a change in rules. Nowhere is this more salient in the current efforts being made by the permanent members of the UN Security Council and the United Nations to fashion some sort of arms trade control regime.

If we are successful in solving some of these puzzles at the systemic level, this puts a premium on explaining and forecasting systemic change. "But changes, and continuities too, are not objective phenomena. Their existence acquires meaning through conceptual formulation and not empirical observation."[3] It becomes very important for observers and analysts of the international arms trade to "pause and ponder what they mean by change and how they would know it when they see it."[4] We must also remember that, as we look for change, not all elements of the system have to change. For example, the control of the export of arms by national governments has been a systemic rule since the late 1930s. But the payoff is change, and it is to this enterprise of defining the international arms trade systems that I now turn.

## Choosing among Competing Approaches and Frameworks

In reviewing the literature to select a framework to be used to describe the various international arms trade systems that have existed, some key questions had to be answered. First was the question of grand systems versus those that are issue specific. If one of the major purposes of this book is the explanation of national actions, either export or acquisition of arms, the framework should be less grand than those that seek to explain the international system as a whole.

Developing an issue-specific international system requires some agreement on boundaries. How do we know the difference between behavior and pressures of the *system* and of the *units* of the system? At the grand level, Waltz has argued that "one must then carefully keep the attributes and interactions of the system's units out of the definition of its structure. If one does not do this, then no systems-level explanations can be given. One cannot even attempt to say how much the system affects the units."[5] In Waltz's view, only when one moves from an "anarchic to a hierarchic realm" does system transformation occur. Since the international system has been basically an anarchic one in the twentieth century, the system can only be changed by a change in distribution of capabilities across units.[6] For the purposes of analyzing international arms trade systems, this approach is inadequate. It does not allow us to explain the very real changes in the international arms trade that have occurred since the 1930s.

The objective is to discern system shifts that affect decisions to import or export arms. A good example is the argument previously mentioned regarding the bipolar nature of the arms trade from the supplier perspective. At the grand level, Waltz might argue that since the United States and the USSR were the major suppliers of the full range of military equipment throughout the post–World War II era, (at least until 1992) the system did not change. But such a conclusion would not give adequate weight to their declining market shares, particularly in certain types of equipment and services critical to the security needs of recipient states. In other words, a framework must be developed that can serve as a guide to explaining systemic effects that are less grand and more issue specific, while avoiding the mere detailing and aggregation of national actions. From the perspective of the national security planners in both supplier and recipient states, how the system constrained and shaped their actions in 1958 was very different from how it did in 1988. If that is true, the goal is to define and explicate arms trade systems that make the transformations clear and useful.

We must also be concerned with the criteria for defining a particular historic system, the classic question of "How do we know it when we see it?" The plausibility of theories, or in this case the definition of a specific arms trade system, is enhanced "if similarities of behavior are observed across realms that are different in substance but similar in structure, and if differences of behavior are observed where realms are similar in substance but different in structure."[7] In the arms-producing and -exporting realm, the substance may indeed look similar. For example, tanks were produced in 1988 much the same as they were in 1958. They were used for similar military purposes, and training and logistics have changed little. But those states acquiring tanks in 1988 did so in a very different manner, due to a systemic shift in the mode of transfer. The 1958 system was marked by grants from the superpowers and a significant set of "strings" attached. Today, wealthy recipients can pay cash, and those who are not so wealthy demand that suppliers come up with offsets in the form of technology transfer, coproduction in the recipient country, and countertrade. This systemic characteristic is not without significant effects, especially when we compare the influence accompanying arms transfers in 1958 with the lack of it in an era of cash and offsets. Conversely, a definition of an arms trade system should allow us to predict similar behavior among suppliers despite drastic differences at the national level. The USSR and the United States had quite different arms-producing and economic systems, but both subscribed to the systemic rule of end-user certificates in order to control the effects of arms transfers.

## Analyzing International Arms Trade Systems Using the Holsti Framework.

The framework chosen to analyze the four historical arms trade systems in this book is that of K. J. Holsti in his text *International Politics*.[8] First, and most important, Holsti is attentive to the major purpose of using international systems, and that is to explain the foreign policy behavior of the units in the system. He is also attentive to system transformation and the development of issue-specific systems. In addition, he eliminates the debate between systems and regimes by including the concept of regime as one of five elements of an international system: boundaries, characteristics of units, structure and stratification, modes of interaction, and regimes. When international relations theorists debate at the grand level, Holsti's scheme may not be central. But for the purpose of explaining why things happen in the international arms trade, his approach is on the mark and quite adaptable to the arms trade issue area. The goal at this point is to outline the components of an international arms transfer system, and then make the empirical observations required to delineate historical systems.

## Boundaries and Environment

Holsti defines the boundaries of a system as that line between interaction and environment "beyond which actions and transactions between the component political units have no effect on environment, and where events or conditions in the environment have no effect on the political units."[9] If arms transfers are to be considered an issue area within the larger international system, how are they to be viewed as distinct from other issues, other types of transactions? How does the arms trade relate to, say, the international political system? The first attempt to utilize the systems approach was that of Harkavy in his 1975 book *The Arms Trade and International Systems*.[10] He constructed a schematic diagram (utilized later in this chapter) that explained arms trade behavior in terms of such political variables as polarity, bloc structure, the role of ideology, and alliance systems. Harkavy isolated arms transfers by creating boundaries so that the linkages to international politics could be better explained. Many other political concepts can be analyzed in this manner; a similar diagram could be constructed to depict the linkage between arms trade behavior and the larger international economic system. What are the dominant modes of economic intercourse in any given historical period? How are critical natural resources controlled and distributed? As the various international arms trade systems are described in this chapter, it will be seen that as the major organizing principles and structure of the international economic system change, the effects of these changes can be linked to the international arms trade behavior of the units in the system.

The same can be said for international security and military systems. At any given time in history, the nature of international conflict and the response of nation-states can be described. In Harkavy's study, for example, he uses the phrase "moderated mood re: total war" to describe the conflict environment in the 1930s, while "zeitgeist of total war" describes the postwar period. During the tight bipolar system of the immediate postwar period, the use of imported arms by clients of the superpowers ran the risk of escalation to the nuclear level. As nuclear parity came to define the international military system, recipients of conventional arms were much more free to use these arms irrespective of the wishes of their donors. A future international security system of ten to twenty nuclear powers would have significant effects on the arms trade behavior of the units. A final component of the international security system linked to the arms trade is military doctrine and strategic thought; to the extent that historical systems are dominated by specific approaches to organizing and fighting wars, the arms acquisition patterns will be effected.

In addition to isolating arms transfer behavior from other different and larger issue-areas, the environment of a system can also have independent effects on unit behavior. For arms transfers, one critical environmental characteristic is technology. Technology is obviously the product of the aggregated efforts of scientists and engineers who reside in and, in the case of military technology, are controlled and financed by national governments. But from the policy-making perspective, decisions are made based on technology as a constant in the short term. Harkavy isolated two criteria that are important: the rate of turnover of generations of weaponry, and changes in actual performance characteristics. He asks, "What might be the implications of these rates of change for the patterns of the arms trade? First, one might expect the arms supplier markets to be more concentrated in times of rapid technological change."[11] For example, the deliverable payload range of an aircraft is a function of technology and shapes how this commodity will be traded. In the current era, much concern has been expressed about many nations acquiring intermediate-to long-range ballistic missiles. The concern arises because of the growth in technology that makes such systems easier to build and acquire.

It is also important to know to what extent armaments production depends on specialized and compartmented technology. How is military production related to production at large in the system? This will significantly affect the ability of governments to control production. And as a final example of how technology can have an impact on unit behavior, units in the system are keenly aware of the state-of-the-art or leading-edge technologies and are driven for prestige reasons to acquire systems containing this technology, often ignoring issues of cost and utility in terms of genuine national security concerns.

*Characteristics of Units*

Holsti's second system component comprises the characteristics of the political units whose interactions form an international system. At the international political system level, this involves observing such things as types of governments and administrations, as well as the methods by which resources of the unit are mobilized to achieve external objectives. For the international arms trade systems being developed here, these general observations can be made specific to the issue area so that, in any given historical system, the characteristics of the arms-trading actors can be used to define the system and explain its transformation.

Of primary importance is knowing which states are suppliers and which are recipients. For each of these two types, a set of rationales exist that tend to dominate any given system. For example, in the 1930s, all of the major arms-trading states agreed to bring under national control the export of military equipment. Until some new suppliers emerged in the late 1970s, it was a characteristic of suppliers that arms would be treated differently than tractors or refrigerators. Other characteristics of suppliers include the percentage of the arms produced in a nation-state that are exported and the degree to which the product is indigenously produced. It is important to define the degree of multinationalization of arms production that is present in the system, and this can only be determined by examining the units themselves. For recipients, indicators of modernity are important in explaining varying levels of technology acquired in weapons systems. Economic conditions of both suppliers and recipients also help explain arms transfer behavior. For example, there is a clear correlation between the oil price increases of 1973, the resulting vast accumulation of petrodollars, and major shifts in arms-trading patterns.

Research on the arms trade in the postwar period has focused mainly on the national governments as actors, given the fact that industrial firms were tightly controlled in their export of arms, and arms transfers tended to be viewed as high politics. But if the systemic analysis is pushed back to the prewar period, to the so-called merchants-of-death era, the firm as unit of analysis must be addressed. Also, as we push the analysis into the future, it appears that the firm is returning as an important unit to be studied and characterized as part of the system. This is particularly true of those firms that produce dual-use products such as computers, commercial jet engines, and communications equipment, since this type of equipment is much less controlled by governmental actors. In sum, at the systemic level, the typical supplier and recipient—both nation and firm—must be described for a given historical period.

## Structure and Stratification

Holsti's next element is the definition at any given time of the configuration of power or influence or persisting forms of dominant or subordinate relationships. What must be described is the manner in which the world is structured for international trade in arms. Important variables include the number of suppliers and recipients, and market shares by weapons systems and country. Dominance and dependence relationships can be depicted by observing the extent to which supplier-recipient relationships are sole-source, multiple-source, or perhaps cross-bloc in those international political systems dominated by a bloc structure. Attention must also be paid to the military potential being traded, since this evidence will serve to answer questions concerning the diffusion of military capability that may account for transformation of the military/security system.

## Modes of Interaction

At the grand international system level, typical analyses address the dominance of certain instruments of foreign policy during a particular historical period. For example, nuclear saber-rattling as an instrument of policy diminished significantly as the USSR and the United States reached parity, and it appears to have disappeared altogether in today's post-cold war system. In an arms transfer system, two modes of interaction are important: how the arms are paid for, and the production arrangements used. Modes of payment can include outright gifts and grant aid, credit, government-to-government versus commercial channels, barter, and offset. It will be seen that this dimension evolves from a predominance of commercial modes in the 1920–40 period to grant aid in the 1950s and 1960s, cash sales in the 1970s, and credits and offsets in the 1980–92 period of limited funds held by potential recipients. Each of these shifts may have their determinants at the national level of analysis. However, once in place, the cumulative effect is a change in the working of the system.

The same can be said for the evolution of production levels as part of international arms trade. The tight bipolar period was dominated by off-the-shelf acquisitions by recipients in the developing world, given their limited economic, political, and technological capacities. This was not true of U.S. exports to a recovering Western Europe, however, as the primary purpose of this aid was to get European arms industries on their feet as soon as possible. This trade was marked by a high degree of licensed and offshore production. It is important, therefore, to analyze the dominant modes of production used in systems being traded—off-the-shelf, licensed production, coproduction, codevelopment, and technology transfer. As with mode of payment, explanations for a preference for a particular type of production scheme on the part of developing countries may very well lie at the national level. For example, the extent to which developing countries now demand offsets as part of an arms deal varies based on national characteristics, capabilities, and previous experience. However, the fact that it now dominates the system is attributable in no small part to imitative behavior based on observations of changing international norms.

There are other characteristics of any arms transfer system that must be addressed under the heading of interaction. For example, Harkavy points out that, in the interwar years, there was a free flow of scientific and technical experts knowledgeable about the development and production of armaments. This flow was indicative of a system in which arms were traded more freely, in stark contrast to the several postwar systems, where arms production expertise was tightly restricted to national boundaries. Also, each system will vary on the dimension of the multinationalization of the corporate arms business. How are arms manufacturers interconnected in terms of business arrangements and codevelopment and coproduction projects? What is the balance between the power of the firm and that of national governments? Interconnected arms firms were a dominant characteristic of the interwar system, and they appear to be emerging as part of the future arms-trading system.

## Regimes, Norms, and Rules of the System

The fifth component of Holsti's systemic approach is the regime. "Interactions and processes in most systems are regulated or governed by *explicit or implicit rules or customs,* the major assumptions or values upon which all relations are based."[12] Regimes are further defined as the "norms, rules, regulations, and decision-making procedures governments develop or use to regulate various types of transactions and/or issue areas."[13] As examples, Holsti anchors the successful end of the regime spectrum with the International Civil Aviation Organization, while at the unsuccessful end of the spectrum he puts the international regime of human rights. In between, he cites the General Agreement on Trade and Tariffs (GATT) and the nuclear nonproliferation regime as fairly successful regimes.

Before outlining what an arms transfer regime might look like, the larger debate on regimes must be briefly addressed. First, as to the question of system versus regime as an organizing construct, regimes are assumed to be part of systems.[14] As for definitions, Holsti subscribes to the consensus definition in Krasner's compendium of regime research:

> Regimes can be defined as sets of implicit or explicit principles, norms, rules, and decisionmaking procedures around which actors' expectations converge in a given area of international relations.[15]

It is also clear that "regimes must be understood as something more than temporary arrangements that change with every shift in power or interests."[16] Simply put, the balance-of-power concept is not a regime. This idea looms important, because an arms transfer regime will be a mix of economics, technology, and national security. As Jervis points out, security regimes are not regimes if the patterned behavior can be explained by "immediate self-interest."[17] This is particularly a problem for arms transfers, since states may refrain from transferring a particularly nasty weapon not because of regime rules but out of fear that the weapon will come back to harm the supplier. In short, it must be demonstrated that national behavior is influenced by the regime.

Puchala and Hopkins put forth five major features of a regime that seem to be relevant and to provide the basis for the construction of comparative arms transfer regimes. First, a regime has an attitudinal component consisting of understandings, expectations, and convictions. Second, a regime contains tenets concerning appropriate procedures for making decisions. Third, there are major principles and norms that prescribe orthodox and proscribe deviant behavior.

Further, there are hierarchies among principles and the prospects for norm enforcement. Fourth, there is a well-defined set of elites who are the practical actors within the regime: governments, and their individual bureaucracies and units that may interact. Finally, wherever there is regularity in behavior, some kinds of principles, norms, or rules must exist to account for it. These patterns may be caused by a powerful actor or an oligarchy instead of voluntary consensus, but it remains a regime nevertheless.[18]

As pointed out earlier, arms transfers are part high politics, part economics and trade, and part military and national security. A brief examination of previous research on three regimes that cover these areas provides some insights into viewing the patterned arms transfer behavior of several historical periods as regimes. Caldwell examines U.S.-Soviet relations in several issue areas (including strategic military, economic, and crisis management) and concludes that, in some, a set of rules and norms developed that patterned the behavior of both nations. For example, in his "cold war nuclear deterrence regime," the first rule is that "each superpower should be able to attack the other superpower with nuclear weapons sufficient to inflict unacceptable damage." Another is that "control of nuclear weapons should not be turned over to the allies of the superpowers."[19] In the economic area, a regime did not develop, whereas in crisis management a set of rules developed to which both countries adhered. Examples of rules in this latter regime include that "communication by deed is more effective than verbal or narrative communication," and that "bargaining is largely tacit."[20] Caldwell's work and that of others who have treated aspects of U.S.-Soviet behavior as a regime[21] provide the impetus to view arms transfers as a regime.

As for the trade issue area, Crawford and Lenway analyzed Western collaboration on East-West trade as a regime.[22] This example is very relevant for the arms transfer issue, because both economic and security issues have been involved in restricting exports of military goods to the Eastern bloc nations. Crawford and Lenway identify three decision modes—compliance, compromise, and problem solving—each of which have a different impact on stable cooperation that they term a regime. These decision modes are a function of power relations that structure interaction among states, goals, and policy instruments within a specific issue area, and the way information is generated and disseminated about the power structure, the goals of others, and the possible consequences of decision. One finding from Crawford and Lenway's examination over time of the actions of Western European states and the United States regarding issues such as technology transfer, credit, and energy reveals that "the relative decline of U.S. economic power, combined with Europe's conflicting goals and interests, partially explains the move from compliance to compromise"[23]—in effect, a regime transformation. The arms transfer issue has a similar structure, with economic objectives often clashing with security goals.

A final example of regime research relevant to analyzing arms transfers in terms of regimes is that of nuclear proliferation.[24] Regimes possess a number of key attributes, including norms that encapsulate their purpose; rules to operationalize these norms; agreements, procedures, and institutions to implement the rules; and degrees of universality of membership. Simpson identified the ethical norm of this regime as one that views nuclear weapons as devices of mass and indiscriminate destruction with unknowable aftereffects, making both them and their use unacceptable to the international community. Politically, there are three conflicting norms. This regime suggests a redistribution of power away from the existing nuclear weapon states through disarmament, whereas the opposite effect could occur if the current situation is frozen in place. The regime also has a third norm, that of reciprocity of benefits and sacrifices within the regime. As for basic rules, the regime calls for those states possessing nuclear weapons to dispose of them and for those without to eschew the attempt to gain access to nuclear weapons. Specific agreements, procedures, and institutions have been set up as part of the regime. There have been several nuclear arms control agreements signed, including the Non-Proliferation Treaty of 1968. The International Atomic Energy Agency was established to monitor the disposition of all fissile materials in peaceful nuclear energy programs and to monitor trade in nuclear materials and technology. In the mid-1970s, further procedures were added to the regime by strengthening safeguards and restricting supplies of nuclear technology and materials. This included the development of several "trigger lists" of materials that suppliers agreed should be the object of strict export restrictions. The nonproliferation regime is often used as a model for a conventional arms trade control regime, with the usual conclusion being that little in the way of rules, procedures, and institutions has evolved for the latter because no ethical norm exists for conventional weapons. This will become particularly clear when the systems in this book are described. However, if the concept of regime is expanded beyond a *control* regime to an arms *trade* regime, it will be seen that such concepts as regime norms and particularly rules are useful in explaining the larger arms trade system.

What these three examples point out is that arraying the evidence on international arms trade into the regime construct has great utility, especially for explaining system transformation. Each historical arms trade regime depicted in the chapters that follow has a different set of principles, norms, rules, and decision-making procedures that explain patterned behavior. However, at the end of chapter 1, a series of questions and propositions were generated based on the literature of the past twenty-five years. We can now use these questions and propositions as a source for regime norms and rules, and to develop for each regime a common set of questions to be asked that can serve as a basis for comparison and explanation of system transformation. For example, since the commodities being traded are military and lethal in nature, one rule or norm to be established for each regime is the effect of the arms trade on regional conflicts or conflict between the major powers. It should be remembered that conventional arms trade first surfaced as a policy problem when the "merchants of death" were thought responsible for World War I and a host of other conflicts.

For each regime, it must be ascertained how the actors in the system perceive the effect of arms transfers on international and intranational conflict, the foreign policy behavior of both supplier and recipient states, and the economic welfare of both suppliers and recipients. What are the norms regarding the extent of national and international control of the exporting of defense production? Are there certain military capabilities in the system that are not exported? What are the dominant demand factors from the perspective of the recipients? It will be argued that the answer to these and other questions will produce a set of "explicit principles, norms, rules, and decisionmaking procedures around which actors' expectations converge"[25] in a given area of international relations. Combined with evidence on the other four elements of a system—boundaries, characteristics of units, structure, and modes of interaction—historical arms trade systems will be developed and explained. These systems are then compared so that the transformation from one to the other may be clarified.

To summarize, the examination of historical arms trade systems will address the following questions for each period.

## VARIABLES FOR EACH HISTORICAL ARMS TRADE SYSTEM

I. Boundaries and environment
   A. Characteristics and structure of larger international systems
      1. Political system (polarity, bloc structure, alliance systems and distribution of power)
      2. Economic system (trade, resources, dominant forms of economic relations, private versus public manufacture and export of arms).
      3. Military system (nature of international conflict, military doctrine)
   B. Technology
      1. Turnover of generations of key weapons systems
      2. Performance characteristics of key weapons systems
      3. Relative importance of specialized versus dual-use production
      4. Prestige, state-of-the-art technologies
II. Characteristics of units (nation-states and firms)
   A. Typical supplier and recipient rationales
   B. Extent of national control over export of armaments
   C. Level of indigenous versus multinational production
   D. Ratio of national production to export
   E. Relationship between economic capabilities and the production, supply, and acquisition of armaments
III. Structure and stratification
   A. Number and size of suppliers and recipients

B. Market shares
  C. Sole- versus multiple-source suppliers
  D. Military potential of exports
IV. Modes of interaction
  A. Hierarchy of modes of payment (grant aid, credit, cash, government-to government, commercial, barter, offsets)
  B. Hierarchy of modes of production involved in arms trade (off-the-shelf, co-assembly, licensing, coproduction, codevelopment, technology transfer)
V. Regime
  A. Principles, norms, rules, and decision-making procedures around which actors' expectations converge
  B. Relationship between arms transfers and other factors
    1. International and intranational conflict
    2. The conduct of foreign policy and diplomacy
    3. Economic welfare of supplier and recipient
  C. Magnitude of control of the arms trade present in the system
    1. Multilateral
    2. Unilateral

# 4
# The Interwar Arms Trade System

Harkavy's *The Arms Trade and International Systems*[1] remains the definitive work on the interwar arms trade system for several reasons. First, it was his doctoral dissertation and makes extensive use of primary sources. For example, most of the data on arms deals of this period came from an exhaustive survey of U.S. military attaché reports from embassies abroad, found in the Modern Military Division of the National Archives in Washington. Very few analyses of any arms transfer period have made as much use of this type of primary historical document. Secondly, as mentioned previously, Harkavy was the first and only one to apply modern social science methods to the interwar period. His explicit use of the systems approach enhances the ability to apply his findings to the schema being developed in this chapter. The books written during and about that era tend to be journalistic and normative in character; many were used and are referred to in the writing of this book and were very useful in providing overviews of the environment that provided the backdrop for the arms trade patterns. But Harkavy alone has produced the systematic assessment that can be fruitfully used in an effort such as this, an attempt to construct and compare the several historical arms transfer systems that have influenced varying behaviors at the nation-state level. Unless otherwise cited, what follows is a summary of his findings adapted to the five-element systemic construct developed by Holsti and described in chapter 3.

## Boundaries and Environment

The first question that must be addressed for this and all historical periods is that of time boundaries. Harkavy chose the period 1930–40 for several reasons. First, attention to the arms trade as an issue commenced soon after World War I, but the international political system was in a state of flux. It was not until 1930, for example, that it was clear that Germany had returned to the international system. Also, by that time the fledgling efforts at doing something about the "evil" of the arms trade were in full bloom. In short, enough time had elapsed from World War I that its effects could be somewhat isolated. Time boundaries are somewhat system dependent, but in Harkavy's study he was relying heavily on empirical data and had to make some choices based on the overall diplomatic history of the interwar years.

This period is sometimes referred to as the "merchants of death" era, referring to the lack of national controls on private arms manufacturers exporting their wares without necessarily considering these exports' economic, political, and military effects. Harkavy's review of the history of national controls on arms exports revealed that an almost totally laissez-faire emphasis in the Middle Ages gave way to some licensing in the period of the Reformation and in the Age of Mercantilism. But throughout the nineteenth century and on into the mid-1930s, national controls were minimal. Although this might demand that the entire period from 1800 to 1935 be treated as a system, too many other variations in the system render such time boundaries inappropriate for the purposes of this study. Since we know that World War II will cause a serious upheaval and transformation of the arms transfer system, it makes more sense to isolate the latter part of the merchants-of-death era.

One of the purposes of Harkavy's study is to explain the arms transfer system as distinct from the larger economic and political forces at work. His diagram see (figure 4–1), truncated to isolate just the interwar system, summarizes the environment and those factors that affect arms. The first set of factors concerns polarity, bloc structure, alliance systems, and distribution of power—in short, the larger international political system. After reviewing contemporary assessments and characterizations of this period, Harkavy describes it as one with relatively dispersed power centers and partial bloc multipolarity. As for the role of ideology as an environmental factor, he notes that, in the 1920s, the primary locus of ideological conflict was between the new Bolshevik regime and the West. By the 1930s, however, the growth of fascism in Italy and Germany had resulted in the tripolarization of international ideological conflict, gradually pulling in the arms trade as part of the pattern.[3] Competing with this realpolitik behavior of states was the larger idealism embedded in the various efforts to bring about peace and disarmament. Most states were forced to get involved in these efforts, at least in the form of responding to the various suggestions made to control national arms production, scuttle battleships, and lower spending on defense. Despite these larger political systemic shifts, Harkavy makes a very strong point that the arms-trading system remained unaffected by them, an "atavism" or remnant of a past era of less than total war. Arms were not traded in accordance with the larger political system, in the form of either the realpolitik behavior that pointed to another major war or the idealistic international peace and security schemes encouraged by the League of Nations. Here we see the utility of using boundaries to separate issue areas from their larger context.

When Harkavy shifts to the larger economic system, he evaluates the linkage between arms transfers and two factors: the dominant forms of

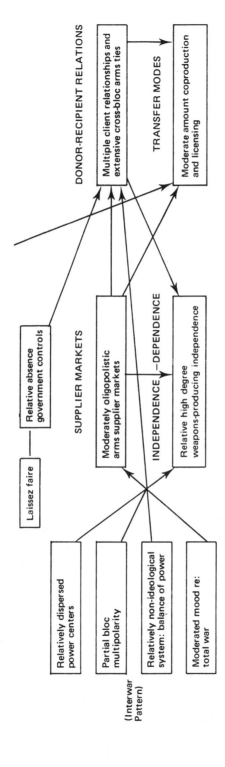

**Figure 4-1. Interwar Arms Trade Relationships**

Source: Harkavy, *The Arms Trade and International Systems.*

economic relations prevailing in given periods, and private versus public manufacture and export of arms. It is puzzling that although the heyday of laissez-faire economics had passed in the early twentieth century, the arms trade of the 1930s was still being practiced in this mode. Again, this gives rise to the notion that this issue area is bounded in a manner that demonstrates that arms trade is not purely economic trade. As for the private-versus-public question, the furor of the post–World War I period had as one of its goals controlling the manufacture of arms as a method of bringing peace to the world. The League of Nations held very visible conferences to this end. Several noticeable scandals involving private arms manufacturers, combined with the anticapitalistic sentiments generated by the Great Depression, added to the pressure for nationalization and reduction of armaments production. But very little effect was seen on arms production or arms trade, save for the monitoring and licensing of arms exports, which by the late 1930s had become a reality for most major governments.

What about the military doctrine and strategies that drove nations to acquire military capability? What was the nature of conflict in this period? Can this knowledge help explain the arms transfer patterns? It was a time of great creativity and doctrinal development in regard to warfare. The integration of the tank and combat aircraft into the inventories of the major powers sparked extensive revisions to theories of armed conflict. There is little doubt that arms acquisitions were related in a major way to the doctrines and modes of warfare developed in this period.

Military technology, as an environmental factor, was marked by a rapid turnover of generations of weaponry, "the time span between the introduction of identifiably qualitative advances in 'state of the art' systems."[4] Harkavy's search of the various compendia of weapons systems from this period reveals a fairly rapid turnover, especially in aircraft and tanks and when compared to the postwar period. This turnover rate is explained in terms of shorter lead times and lower unit costs, as well as the fact that until war became imminent in the late 1930s, production seemed to focus on the rapid development of a few prototypes versus a larger stockpiling of inventories in anticipation of actual use in combat. It also must be remembered that the tanks, aircraft, and ships of the interwar period were a relatively new concept, and experimentation would account for much of the turnover. Harkavy notes a decline in generational turnover as World War II approached and states began to stock up for the conflict.[5] This is a phenomenon that also occurred in later periods, especially during the U.S. involvement in Vietnam: research and development budgets declined significantly when a scarce-resources situation prevailed in the late 1960s, but equipment still needed to be procured for the forces engaged in combat, both U.S. and South Vietnamese.

Another element of the technological environment is the relative importance of specialized military materials and skills needed for production of weapons systems versus standard commercial materials and skills. The "business end" of most systems was the warhead. In the interwar period, a critical material for producing military capability in general was the chemicals needed for the explosives in these warheads. The major chemical cartels, such as Nobel and DuPont, were the targets of much of the criticism of the arms trade in the interwar years. But these cartels did much more than supply the raw materials for military explosives. Their wide range of civilian products seriously hampered efforts to connect these firms to destabilizing arms production and attempts to institute international and national controls. Another facet of this phenomenon was the generally low technology being used in developing aircraft and tanks; production was much more a function of investment and large quantities of manpower applied to major production and assembly operations. Had military production been more specialized, and the raw materials needed for production limited to fewer suppliers, the arms trade system may well have looked different.

This technological environment can be related to the arms trade system in several other ways. First, as the qualitative arms race ensues, not everyone can continue to compete in terms of production. Harkavy finds, for example, that by the late 1930s the previous capability of nations such as France, the Netherlands, and Czechoslovakia to produce first-line equipment had begun to fade.[6] Also, rapid generational evolution of systems may result in pressures to sell the older systems. It also can create pressure to procure the "latest" system for prestige reasons.

## Characteristics of Units

The units in the interwar arms trade system were national governments, arms firms controlled by national governments, and private firms that were basically free to trade in armaments across national boundaries. The typical supplier of arms was concerned first with the economics of the deal, with very few examples of arms transfers being used for political or diplomatic purposes. The entire world was the marketplace, as the growing tripolarization did not affect arms trade until late in the period. Until the mid-1930s, national control over arms exports was minimal.

There was a definite hierarchy of producers of military equipment, as indicated in table 4–1. Level 1 dependence is defined as total independence and level 2 as near-total independence, whereas level 6 is total dependence. For aircraft, it can be seen that Germany, Italy, Britain, and the United States were fully independent, while France, the Netherlands, and Poland were close. It should also be noted that the USSR and Japan were still dependent on imports and were busy copying and reverse engineering, a strategy that would pay off in indigenous production during World War II.[7]

The correlation between producers/suppliers and their economic size is revealed by comparing table 4–1 with statistics on arms exports. Produced in

## Table 4-1
### Summary of Interwar Dependence Levels by Weapons System

| Weapons System | Level | Countries |
|---|---|---|
| Aircraft | 1 | Germany, United Kingdom, United States |
| | 2 | France, Poland, Netherlands |
| | 3 | USSR, Japan, Sweden, Czechoslovakia |
| | 4 | Canada, Spain, Yugoslavia, Belgium, South Africa, Switzerland, Denmark, Norway, Greece, Romania Thailand |
| | 5 | China, Australia, Argentina, Brazil, Hungary, Mexico Turkey, Austria, Chile, Bulgaria, Finland Lithuania, Portugal |
| | 6 | All remaining nations |
| Naval | 1 | United States, Germany, United Kingdom, France, Italy, Sweden, Netherlands, Denmark, Norway |
| | 2 | Japan |
| | 3 | USSR, Spain |
| | 4 | Australia, Argentina, Yugoslavia |
| | 5 | Brazil, Turkey, Finland, Portugal |
| | 6 | All remaining nations |
| Armor | 1 | United States, Germany, United Kingdom, France, Italy, Czechoslovakia |
| | 2 | Japan |
| | 3 | USSR, Sweden, South Africa |
| | 4 | Poland, Switzerland, Hungary, Austria |
| | 5 | Netherlands, Yugoslavia, Belgium, Romania |
| | 6 | All remaining nations |
| Small arms | 1 | United States, Germany, United Kingdom, USSR, France, Italy, Sweden, Belgium, Czechoslovakia, Switzerland, Denmark |
| | 2 | Japan |
| | 3 | Australia, Poland, Netherlands, Yugoslavia |
| | 4 | Canada, Spain, Argentina, Austria, Finland, Portugal, Romania, Iran |
| | 5 | China, South Africa, Hungary, Mexico, Turkey, Norway, New Zealand, Bulgaria |
| | 6 | All remaining nations |

Source: Harkavy, *The Arms Trade and International Systems,* p. 192.

table 4-2 are Harkavy's data on combat aircraft,[8] which show that arms development and production were not necessarily seen as an "essential hallmark of national power. . . . Russia, Japan, Turkey, and to a lesser extent, Austria-Hungary, in addition to Italy, were then essentially arms dependent upon the corporations of France, Britain, Germany and the U.S. with smaller supply centers in Belgium, Sweden, Switzerland and Denmark (Fabrique Nationale, Bofors, Oerlikon, and Madsen) playing a surprisingly important role."[9]

There was a definite tendency for arms firms of this period to set up subsidiaries around the globe, affording a form of technology transfer on a continuing basis, with minimal blockage from the governments of the supplier

### Table 4-2
### Combat Aircraft Exported, 1930-39

| Rank | Country | Number | Percentage of Total | National Income (billions of dollars) |
|---|---|---|---|---|
| 1 | United States | 3,218 | 22.8 | 67.4 |
| 2 | United Kingdom | 2,435 | 17.3 | 22.1 |
| 3 | France | 2,204 | 15.6 | 10.8 |
| 4 | Italy | 1,786 | 12.7 | 5.8 |
| 5 | Germany | 1,336 | 9.5 | 23 |
| 6 | USSR | 784 | 5.6 | 17.9 |
| 7 | Netherlands | 429 | 3.0 | 2.7 |
| 8 | Poland | 306 | 2.2 | — |
| 9 | Other | 1,614 | 11.3 | — |

Other Suppliers: Belgium, Canada, Czechoslovakia, Sweden, Switzerland, Japan, Denmark, Yugoslavia, Mexico

Source: Harkavy, *The Arms Trade and International Systems*, pp. 71, 91.

corporations. British shipbuilders had facilities in Finland, Spain, and Italy. The Germans used Swedish subsidiaries for arms development until the foreboding political developments of the mid-1930s brought it to a close.[10] It was not until this latter development that arms firms began to be more circumspect with their knowledge, fostered in part by increased cooperation and eventually control of arms exports by national governments.

## Structure and Stratification

### Suppliers and Recipients

One of the major ways to compare systems and to describe their transformation is by first determining the number of suppliers and recipients. In the interwar system, Harkavy identified nineteen countries that exported aircraft, nine that exported tanks, and nine that exported warships. The level of market concentration, as measured by the percentage of items exported for the highest two suppliers, varied from 40 percent for aircraft to 54 percent for tanks and 77 percent for warships. Harkavy also identified fifty-five recipients that produced very little in the way of indigenous armaments and relied on imports to equip their national forces. It should be remembered that a very large number of what were to become recipients in the postwar system were colonies in the interwar period; none of the intracolonial arms transfers is considered as arms trade in his book.

### Market Shares

Market shares in 1930 are detailed in table 4-3 for tanks, warships, and combat aircraft. The League of Nations put forth a major effort to collect and publish

Table 4-3
Market Shares for Tanks, Warships, and Combat Aircraft, 1930

| | Tanks | | | | Warships | | |
|---|---|---|---|---|---|---|---|
| Rank | Country | Number | Percentage of Total | Rank | Country | Number | Percentage of Total |
| 1 | France | 1,091 | 27.9 | 1 | United Kingdom | 76 | 58.9 |
| 2 | United Kingdom | 1,017 | 26.1 | 2 | Italy | 23 | 17.8 |
| 3 | United States | 574 | 14.7 | 3 | France | 13 | 10.1 |
| 4 | Italy | 424 | 10.9 | 4 | Spain | 5 | 3.9 |
| 5 | Czechoslovakia | 270 | 6.9 | 5 | Japan | 5 | 3.9 |
| 6 | USSR | 220 | 5.6 | 6 | United States | 3 | 2.3 |
| 7 | Germany | 160 | 4.1 | | | | |
| 8 | Sweden | 122 | 3.1 | | | | |
| 9 | Poland | 26 | 0.7 | | | | |

Combat Aircraft

| Rank | Country | Number | Percentage of Total |
|---|---|---|---|
| 1 | United States | 3,218 | 22.8 |
| 2 | United Kingdom | 2,435 | 17.3 |
| 3 | France | 2,204 | 15.6 |
| 4 | Italy | 1,786 | 12.7 |
| 5 | Germany | 1,336 | 9.5 |
| 6 | USSR | 784 | 5.6 |
| 7 | Netherlands | 429 | 3.0 |
| 8 | Poland | 306 | 2.2 |

Source: Harkavy, *The Arms Trade and International Systems*, pp. 61, 69, 74.

arms export data; table 4-4 gives an additional look at market shares based on dollar amounts rather than equipment types. It can be seen that countries such as Czechoslovakia, with its historically powerful Skoda firm, appear as major actors despite their lower rankings for major types of military equipment. Conversely, the United States appears in a somewhat lower position, perhaps due to its skewed investment in the aircraft industry.[11]

Harkavy also makes the point that within this period, market share shifted to the Axis or revisionist powers.[12] When comparing market share shifts from the 1930-34 period to the 1935-40 period, Harkavy finds that in general the United States, the United Kingdom, and France experienced declines in most categories of weapons exports, while the revisionist powers increased their share. This infers that in the latter part of the period the market began to be structured more in line with emerging foreign policies designed to exert influence and create a revised international political system, and less with that of a laissez-faire economic environment.[13]

*Supplier-Recipient Dependencies*

The dependencies in the system were noted in table 4-1. Harkavy rated each of the countries in that period on a total independence (1) to total dependence (6) scale across four types of weapons systems (aircraft, naval, armor, and small arms). He then correlated these rankings with national income. Harkavy found that most or all nations high in gross national income and per capita income "have been capable of developing and producing their own weapons, although in some cases maintaining a degree of dependence by choice. On the other end of the spectrum, those nations with both relatively low GNP and lower per capita GNP have not been able to produce their own arms even with the aid of licensing and copying."[14]

As for the recipients of this era, Harkavy arrayed his data in terms of three acquisition styles: sole supplier, predominant supplier, and multiple supplier. As table 4-5 indicates, the predominance of multiple-supplier relationships indicates an environment in which acquisition rationales were not very dependent on the ideological and political structure of the larger international system. His data on acquisitions by China are indicative of the freedom that recipients had to shop for the best economic deal in light of their military needs.[15]

When describing the sole-supplier, predominant-supplier, and multiple-supplier structure of the system, it is important to remember that during this period the modern political science concepts of dominance, dependence, and independence are less meaningful when applied to the arms trade than in later periods. Private arms firms are the basic unit of analysis in this system. It is their behavior that needs to be explained, not that of governments.[16] Even Harkavy's treatise, though, does not attempt to do that. Therefore, comparisons using data on supplier-recipient structure can be misleading. It may suffice to conclude the comparison by stating that an international arms trade structure dominated by

Table 4-4
Export of Arms and Ammunition of Principal Exporters, 1929-38
*(in millions of former gold $)*

|  | 1929 | 1930 | 1931 | 1932 | 1933 | 1934 | 1935 | 1936 | 1937 | 1938 | Average | Percentage of Total |
|---|---|---|---|---|---|---|---|---|---|---|---|---|
| United Kingdom | 21.8 | 17.0 | 13.4 | 10.1 | 10.1 | 8.5 | 9.6 | 9.7 | 11.9 | 17.1 | 12.9 | 25.0 |
| Czechoslovakia | 3.2 | 5.3 | 3.9 | 1.4 | 3.2 | 8.7 | 10.5 | 7.7 | 7.2 | 14.3 | 6.5 | 12.6 |
| Germany | 7.5 | 6.5 | 5.2 | 3.8 | 4.1 | 3.3 | 2.5 | 5.3 | 9.0 | 8.4 | 5.6 | 10.9 |
| France | 9.4 | 7.0 | 2.7 | 9.3 | 8.5 | 8.0 | 6.6 | 10.6 | 6.7 | 6.8 | 5.5 | 10.7 |
| United States | 10.7 | 6.5 | 3.9 | 2.9 | 3.2 | 3.7 | 3.3 | 4.3 | 5.6 | 6.9 | 5.1 | 9.9 |
| Sweden | 3.0 | 4.3 | 3.7 | 3.7 | 3.4 | 3.7 | 3.3 | 2.9 | 5.7 | 7.8 | 4.2 | 8.1 |
| Belgium | 3.0 | 2.5 | 1.5 | 1.5 | 1.4 | 2.1 | 2.5 | 2.4 | 3.2 | 6.1 | 2.6 | 5.0 |
| Italy | 3.7 | 3.8 | 2.2 | 0.6 | 1.5 | 1.6 | 0.7 | 0.7 | 1.6 | 2.2 | 1.9 | 3.7 |
| Switzerland | 0.5 | 1.0 | 1.1 | 0.7 | 1.1 | 0.8 | 0.4 | 2.4 | 5.0 | 5.8 | 1.9 | 3.7 |
| Others | 9.6 | 8.3 | 2.8 | 4.6 | 5.6 | 4.6 | 3.7 | 3.9 | 4.5 | 6.5 | 5.4 | 10.4 |
| Total | 72.4 | 62.2 | 40.4 | 38.6 | 42.1 | 45.0 | 43.1 | 49.9 | 60.4 | 81.9 | 51.6 | 100.0 |

Source: Nokhim M. Sloutzki, *The World Armaments Race, 1919-1939* (Geneva: Geneva Research Center, 1941), p. 71.

### Table 4-5
### Interwar Supplier-Recipient Patterns by Acquisition Style

| Sole Supplier | | Predominant Supplier | | Multiple Supplier |
|---|---|---|---|---|
| *West bloc* | | *West bloc* | | *West bloc* |
| Cuba | US | Australia | UK | Guatemala |
| Egypt | UK | Canada | UK | Loyalist Spain |
| Haiti | US | Costa Rica | Fr | South Africa |
| Honduras | US | Dominican Republic | US | Belgium |
| Saudi Arabia | UK | Eire | UK | Czechoslovakia |
| New Zealand | UK | Estonia | UK | |
| | | Mexico | US | *Axis bloc* |
| *Axis bloc* | | Poland | Fr | None |
| Albania | It | | | |
| | | *West bloc with* | | *Cross-bloc* |
| | | *cross-bloc ties* | | Argentina |
| | | Colombia | US | Austria |
| | | Greece | UK | Bolivia |
| | | Iraq | UK | Brazil |
| | | Latvia | UK | Chile |
| | | Nicaragua | US | China |
| | | Portugal | UK | Denmark |
| | | | | Ethiopia |
| | | *Axis bloc with* | | Finland |
| | | *cross-bloc ties* | | Iran |
| | | Afghan | It | Lithuania |
| | | Bulgaria | Ger | Norway |
| | | Ecuador | It | Peru |
| | | Hungary | It | Rumania |
| | | Paraguay | It | El Salvador |
| | | | | Sweden |
| | | *Axis bloc* | | Switzerland |
| | | Nationalist Spain | It | Thailand |
| | | | | Turkey |
| | | | | Uruguay |
| | | | | Venezuela |
| | | | | Yugoslavia |
| | | | | Netherlands |

Source: Harkavy, *The Arms Trade and International Systems*, p. 115
Note: Identity of Sole or Predominant Supplier indicated in columns to right of country, where appropriate.

large, uncontrolled arms firms operating on market principles is qualitatively different. Foreign policies generated in such a system will be fundamentally different, at least insofar they employ arms transfers as an instrument of policy.

### Regional Distribution

Harkavy's study did not produce data on the regional distribution of arms transfers. There is a general lack of information in the literature not only on the regional focus but also on the rationales and other characteristics of recipients in the interwar system.

## Proliferation of Military Potential

The rapid generational turnover of equipment produced and traded during this period led to a rapid improvement in the performance characteristics of the weapons systems. Harkavy produced data in this area for aircraft and tanks. Although the assigning of capability based on weapons characteristics is complex,[17] we can fruitfully look at these data (table 4-6) and estimate the effect on the arms trade system. For example, there was a doubling of fighter aircraft range from 1930 to 1939. Similarly, the range of speed of tanks increased from 15-28 mph in 1930 to 20-40 mph by 1939.[18] Those countries expecting war in Europe and in Asia were logical customers, whereas developing countries faced with internal war could be seen as less likely to buy, especially given the price tag. Harkavy also found that the trade of this era was in systems more modern than would be found in the immediate postwar system.[19]

## Modes of Interaction

Harkavy's research produced clear findings regarding the modes of transfer that dominated the system. First, he could find very few instances of grant aid. This fits with the apolitical nature of the trade, at least for most of the period. Most of the arms deals were based on the financial system of the period (mainly cash sales and some credit). It was the private firm, working with banks, that determined the mode of payment. As for how arms technology and production were diffused throughout the system, several things are clear. First, coproduction and codevelopment, modes that became increasingly important after World War II, were unheard of in the interwar period. The licensing of arms production, however, was a dominant mode of transfer during the interwar period. Licensing was most frequent in those countries most dependent on

### Table 4-6
Modernity Indices—Aggregate Comparison of Interwar and Postwar Systems
*(percentages)*

|  | Interwar Years | | | Postwar Years | | |
| --- | --- | --- | --- | --- | --- | --- |
|  | Under 5 | 5-10 | Over 10 | Under 5 | 5-10 | Under 10 |
| Combat aircraft | 85.8 | 11.0 | 3.2 | 16.9 | 50.4 | 32.7 |
| Transport aircraft | 65.5 | 8.8 | 25.4 | 17.1 | 30.5 | 52.4 |
| Trainers | 74.5 | 21.2 | 4.4 | 17.0 | 32.6 | 50.4 |
| Tanks | 74.8 | 14.9 | 10.3 | 13.0 | 27.8 | 59.2 |
| Armored cars | 67.1 | 4.4 | 28.5 | 33.3 | 27.2 | 39.5 |
| Submarines | 98.9 | 1.0 | 0.0 | 27.5 | 20.9 | 51.6 |
| Warships | 93.5 | 2.4 | 4.1 | 38.8 | 18.5 | 42.7 |
| Patrol vessels | 95.6 | 0.9 | 3.5 | 59.4 | 16.3 | 24.3 |

Source: Harkavy, *The Arms Trade and International Systems*, p. 93.

specific weapons systems, and in those weapons systems made up of less sophisticated technologies. Table 4–7 depicts this situation.[20]

The licensing arrangements of the period also tended to reflect the general cross-bloc political arrangements. "Licensing agreements occurred even where there were fairly low levels of internation association, in conjunction with cross-bloc acquisition patterns, and frequently involving relatively underdeveloped nations."[21]

Another unique feature of the interwar system was the highly developed mode of transfer of copying.

> In the interwar period, however, with its laxness of controls and relative freedom of private firms, the art of copying was institutionalized in a different form. Nations short on indigenous weapons development talent were then able to make a practice of buying up one or a few units of given systems, which they then copied or modified. This was done with the knowledge—and apparent resigned tolerance—of the original developers, who in turn were not usually restrained by their own government.[22]

Both Japan and Russia succeeded in obtaining not only copies of the new aircraft models, but often obtained licenses for production also, which would include blueprints and other technical information that eliminated the need for outright copying. In such an environment, "stealing," or blatantly illegal arms transfers, was minimal.

One final type of transfer mode, the retransfer of weapons systems, needs to be mentioned, even though Harkavy concludes that it was not a critical feature of the interwar period. The relatively free trading system meant that retransfer restrictions were minimal and loosely enforced. But retransfers were less prevalent because unit costs were low enough that new equipment could usually be acquired, and because there were fewer developing countries at that time. An additional characteristic of the system, the fact that the major suppliers were essentially independent of each other (that is, there were few licenses and no coproduction arrangements), reduced the incentive to retransfer.

## Regime Norms and Rules

How, then, can this system be characterized in terms of the norms, rules, and decision-making rules around which the actors in the system converged? As indicated at the end of chapter 3, the task at hand is to develop specific answers for the interwar period regarding the four major areas of regime rules and norms: how arms transfer decisions relate to conflict, foreign policy, and the economic welfare of both supplier and recipient, as well as to the rules regarding the control of the arms trade. The rules and norms developed appear under the headings below.

Table 4-7
Transfers Accounted for by Licensing, Interwar Suppliers (percent)

| Supplier Nation | Combat Aircraft | Transport Aircraft | Trainer Aircraft | Tanks | Armored Cars | Submarines | Warships | Patrol Vessels |
|---|---|---|---|---|---|---|---|---|
| Czechoslovakia | 31.4 | 0.0 | 67.1 | 14.8 | 0.0 | 0.0 | 0.0 | 0.0 |
| France | 26.3 | 0.0 | 13.8 | 10.5 | 0.0 | 0.0 | 61.5 | 0.0 |
| Italy | 2.1 | 0.0 | 41.1 | 0.0 | 0.0 | 0.0 | 0.0 | 69.9 |
| Netherlands | 31.7 | 0.0 | 0.0 | 0.0 | 0.0 | 0.0 | 0.0 | 0.0 |
| Sweden | 40.4 | 0.0 | 0.0 | 24.6 | 0.0 | 0.0 | 0.0 | 0.0 |
| United Kingdom | 34.5 | 100.0 | 25.4 | 16.9 | 0.8 | 27.8 | 15.8 | 11.3 |
| United States | 8.5 | 15.2 | 12.0 | 52.3 | 69.4 | 0.0 | 0.0 | 0.0 |
| Poland | 68.6 | 0.0 | 0.0 | 0.0 | 0.0 | 0.0 | 0.0 | 0.0 |
| Germany | 23.8 | 27.6 | 49.5 | 0.0 | 0.0 | 50.0 | 0.0 | 0.0 |
| All Suppliers | 17.5 | 21.0 | 23.8 | 17.0 | 15.7 | 7.6 | 11.8 | 23.4 |

Source: Harkavy, *The Arms Trade and International Systems*, p. 151.

## Conflict

**Arms can be exported to any country, irrespective of ongoing or potential regional and international conflict.** For most of this period, decisions to transfer arms were made independent of ongoing and emerging conflicts. The minimal national controls meant that decisions were made by private firms based on economic considerations. Even late in the period, with World War II only two years away, the United States and others were transferring equipment and technology to Japan and Germany. Indigenous production decisions did reflect these emerging threats, but the arms trade did not.

**Arms transfers are not critical in the development or exacerbation of interstate conflict. Therefore, national actors should not control their exports. To the contrary, maximum and free access to arms supplies should be fostered on an international level to ensure maximum capability to arm in the event of an emergency.** How did the arms trade contribute to the onset, exacerbation, and resolution of armed conflict? It must be remembered that international arms trade became an international public issue in the 1920s expressly because of the perceived linkage between arms transfers and war. Indeed, World War I was often blamed on the "merchants of death." Was this proposition ever proven, in the scientific sense of the word? This is a difficult and contentious issue even in the current time frame, when the study of conflict by social scientists can be considered well developed, and the evidence required to resolve the issue is much more public that was the case in the interwar period. The arms trade data required to answer the question for the interwar period resides with the private firms who dominated the interactions, and despite some availability of these archives,[23] no case studies of that period exist that would allow us to examine the linkage more closely. All we have are the public investigations that did take place regarding the linkage between the profit motives of private arms dealers and peace.

At the international level, these investigations took place in the various international forums convened to deal with the aftermath of World War I. These included the Paris peace conference of 1919 and the Convention for the Control of the Trade in Arms and Ammunition of St. Germain, an international arms traffic conference in 1925, and the Commission for the Regulation of Arms and Implements of War in 1932. Not surprisingly, these deliberations did not resolve the central question: is it the munitions makers and dealers who are causing conflict, or is conflict caused by other factors (such as the human condition and national rivalries), with arms manufacturers as mere suppliers? Wiltz's excellent study of the U.S. Senate Munitions Inquiry of 1934–36 (the Nye committee hearings) concludes that most of the evidence produced at the international level "came from second-hand sources, and none of it established the premises from which conclusions derived. In the climate of opinion prevailing in 1934 few defenders of munitions makers ventured forth."[24] As far as the Nye committee's investigation, Wiltz concludes that one of the most notable achievements "was debunking—inadvertently to be sure—the merchants of death thesis. For some twenty months the committee focused on the thesis, proving in the end that it had limited validity. Thereafter the thesis, a national distraction during a critical period, faded from the American mind. By 1938 hardly anybody bothered with merchants of death."[25] Wiltz also confirmed in a 1958 interview that Nye told him "he never believed munitions makers created tensions and war to promote sales."[26]

But in an important way, the "scientific" evidence was not important. What dominated the mind-sets of the public was the view that private arms manufacturers were in fact responsible for the onset and exacerbation of conflict. Wiltz summarizes the view in the United States, a view that could be expected to be even more prevalent in a Europe that was much closer to the reality of international conflict:

> There was President Franklin D. Roosevelt's statement that "the private and uncontrolled manufacture of arms and munitions and the traffic therein has become a serious source of international discord and strife." If such allegations had been country store or mainstreet gossip that would have been one thing. They were something else when echoed by the President of the United States, the Premier of France, two former Secretaries of State, the League of Nations, *Fortune* Magazine, the *Christian Science Monitor,* members of Congress, the peace movement, leaders of religion, and even the *Wall Street Journal* and the Chicago *Journal of Commerce.* Who could blame people in the mid-thirties for taking seriously this heady business of merchants of death? It seemed so reasonable. Businesses always wished a climate favorable to their merchandise, and if markets did not exist they tried to create them. Why would people selling munitions act differently?[27]

It was this view that created pressure on the decision makers of the system to control the arms trade. And, in the end, that is what they did.[28] As World War II approached, all of the major suppliers agreed to the national control of either the arms trade or production, evidence that at the political level the system recognized that arms trade can create and exacerbate conflict and should be controlled by the entity charged with the prevention and prosecution of conflict, namely, the nation-state. But the internationalization of this norm came late and was not part of the interwar international arms trade system.

*Foreign Policy and Diplomacy*

**Arms transfers patterns do not coincide with the dominant international political structure.** In the realm of foreign policy, arms transfers in the interwar years were hardly what they were to become in the postwar era—"foreign policy writ large."[29] Harkavy's evidence makes it clear that until late in the period, arms transfer patterns were characterized by cross-bloc patterns.

**Arms exports are not used as an instrument of foreign policy, either to gain political influence with recipients, to establish presence in the forms of bases or strategic access, or to deter or punish by the use of arms embargoes.** The use of arms transfers to influence the behavior of recipients was not a feature of this system. It was an age of idealism and legalism at the international level, in which power politics utilizing arms exports as leverage was out of place. The control of exports resided at the corporate level and precluded any such use of this instrument, even if nations had been so inclined. Government personnel from the supplier states, such as military attachés, were stationed in many recipient countries. But most often, they acted as agents for private arms manufacturers, not as representatives of the supplier government with the mission of overseeing an influence attempt. Arms transfers were not used as part of basing or strategic access agreements: "The strategic access afforded the major powers was primarily a function of their colonial possessions, protectorates and mandates."[30] Arms embargoes were not employed. Diplomacy did not, of course, disappear in this period. Rather, the export of arms was not used as an instrument of policy.

*Economic Effects*

**Economic benefits to the citizens of supplier and recipient nation-states from the arms trade is determined by arms manufacturers based on free-enterprise market forces.** What of the economic effects of the arms trade, on both the supplier and recipient? In the postwar period, it can be seen that economics play a crucial role in arms transfers. Supplier states gain economic benefits from arms exports due to longer production runs and lower unit costs. Recipients now look for suppliers to offset arms exports with jobs and technology transfers to bolster their economies and to develop their own indigenous industrial capacity. In short, arms transfers can be best understood as falling "within the scope of activities understood as political economy, i.e., that domain of human activity that is the intersection of the pursuit of political power and purpose with the production of goods and services. These activities are responses to two pervasive systemic imperatives animating state and governmental behavior: order and welfare."[31] Achieving order is accomplished by the state actions regarding conflict and political influence (see rules above). Welfare would affect state decisions regarding arms transfers if costs or benefits accrue from the activity. During the interwar years, the interaction between order and welfare (in other words, the political economy of making and marketing arms) was dominated by the quest for order through traditional means of power and diplomacy. Arms transfer patterns were significantly shaped by economic forces.

## Arms Trade Control

**State actors should ensure that no international control regime is created that would restrict access to armaments that may be needed for defending the national interests of the state.** What controls were there on the national government seeking to export or import weapons systems? First, there was the aforementioned international outcry from the public about the merchants-of-death thesis. Throughout this period, decision makers had to be attentive to this pressure, although it was not until the onset of World War II that behavior changed. The pressure did mean that, despite armament firms operating within their borders and exporting arms that did have impacts on international politics, countries had to attend international conferences, to agree to lofty goals about controlling the trade, and in general, to conform to the international efforts seeking to control conflict through publicity about arms production and trade. The Nye committee hearings in the United States is a good example of this.

It might also be said that regarding controls, the major goal of a national actor in this period was to work to prevent their existence. Since there was little hard evidence that the pacifist arguments were true, and it was hardly likely that all states would cooperate in any international control regime, the most prudent strategy was to ensure that as many sources of armaments as possible were available for acquisition. This certainly was the strategy of Germany, the fledgling Soviet Union, and Japan.

Table 4-8 presents a summary of the arms trade system for the period 1930-40.

**Table 4-8**
**Summary of Interwar Arms Trade System**

| | |
|---|---|
| *Boundaries and Environment* | • 1930-40<br>• Arms trade lags behind larger political shifts as WWII approaches<br>• Arms trade remains laissez-faire despite shift to more national control of economies<br>• Rapid technological change |
| *Characteristics of Units* | • Private arms firms and their subsidiaries dominate system<br>• Trend toward national government involvement as WWII approaches<br>• Multiple-supplier relationships<br>• Acquisition rationales not dependent on ideological and political structure of system |
| *Structure and Stratification* | • West Europe and U.S. suppliers dominate market, with Axis powers increasing market share toward end of period<br>• Many countries can independently produce major weapons systems<br>• Dependence a function of wealth |
| *Modes of Interaction* | • No grant aid<br>• Credit provided by private banks and firms<br>• Licensing dominant mode of transfer<br>• Very little coproduction of codevelopment |

*Table 4-8 continued*

*Regime Norms and Rules*

Conflict
- Arms can be exported to any country irrespective of ongoing or potential regional and international conflict.
- Arms transfers are not critical in the development or exacerbation of interstate conflict. Therefore, national actors should not control their export. To the contrary, maximum and free access to arms supplies should be fostered on an international level, to ensure maximum capability to arm in the event of an emergency.

Foreign Policy and Diplomacy
- Arms transfers do not coincide with the dominant international political structure.
- Arms exports are not used as an instrument of foreign policy, either to gain political influence with recipients, to establish presence in the forms of bases or strategic access, or to deter or punish by the use of arms embargoes.

Economic Effects
- Economic benefits to the citizens of supplier and recipient nation-states from the arms trade are determined by arms manufacturers based on free-enterprise market forces.

Arms Trade Control
- State actors should ensure that no international control regime is created that would restrict access to armaments that may be needed for defending the national interests of the state.

# 5
# The Evolution of Arms Trade Systems in the Postwar Period

## The Postwar Bipolar System, 1946-66

The effects that World War II had on the transformation of the international system are well-known and need little elaboration here. A total war involving the entire system reduced the number of major national actors to two; most of the arms production capacity of the other actors was destroyed. In a sense, the immediate postwar era can be viewed as a quasi-experiment in which two widely different temporal systems could be fruitfully compared, with little controversy that a systemic transformation had occurred. This is not to say that the seeds of some of the changed rules of the postwar regime were not sown prior to the war. Rather, the totality and suddenness of the armed conflict accelerated the changes and created two starkly different systems.

### Boundaries and Environment

As with the previous system, the first question is one of time boundaries. Unlike the interwar period, however, the significant amount of research on the arms trade since the 1970s has seen many attempts at systematic boundary definition. Harkavy's study, the only one to integrate the interwar years in any systematic fashion, has 1968 as its upper boundary. Although this is in no small part because his research was conducted in the 1969-72 time frame, Harkavy also notes in his 1975 foreword that "events in the Middle East in October, 1973 strongly indicated that a new era of conventional warfare and weaponry had dawned, and concomitantly, that the basic nature of the arms trade had also been altered."[1]

Some of the research conducted later in the 1970s and 1980s provides further perspective on temporal boundaries. Canizzo entitled the immediate postwar period "Alliance Building—The Cold War Years" and points to its transformation in the middle to late 1960s, becoming fully transformed after the 1973 OPEC decision to raise oil prices.[2] The 1978 book *Arms Transfers to the*

*Third World: The Military Buildup in Less Industrial Countries* is a compilation of papers from a 1976 conference. The premise for the conference was stated in the preface of the book: "Over the past decade, the diffusion of military technology from the industrial to the less industrial world has increased dramatically. . . . This phenomenon is linked, moreover, to the effects of the quantum jump in weapons technology itself, a development that has transformed dramatically both the conduct of [non-nuclear] warfare and military doctrine, as well as related politico-strategic concepts."[3] The "modern" in Neuman and Harkavy's 1979 text *Arms Transfers in the Modern World* stems from the many changes (outlined in the text) since the dawn of the 1970s. The one systemic chapter, by Mihalka, specifically looks at 1967–76 and compares it with the 1958–66 period.[4] Pierre's very influential 1982 book *The Global Politics of Arms Sales* uses data up to 1978 and constantly refers to the decade of the 1970s as a new system.[5] Finally, the most recent assessment of the postwar arms trade by SIPRI confirms that, on the basis of statistics and "inherent structural elements, the mid-1960s represents the upper limit of this first post-war system."[6] In sum, previous studies of this period reach a consensus on the mid-1960s as the upper temporal boundary of the system.

When this time period is empirically described, the justification for an upper temporal boundary of 1968–70 becomes more clear. Figure 5–1 indicates a

**Figure 5–1. Arms Exports to the Third World, 1951–85 (Major Weapons, 1985 US $m) 3 Year Moving Average**

Source: Brzoska and Ohlson, *Arms Transfers to the Third World: 1971–85*.
Note: Figures represent three-year moving averages of major weapons exports in constant 1985 U.S. millions of dollars.

clear step-level change in arms transfers to the Third World occurring in the late 1960s. ACDA data on conventional arms *deliveries* also supports these boundaries. Catrina has converted ACDA data to constant 1984 dollars, allowing figure 5–2 to depict systemic shifts.

Because Harkavy explicitly compared the interwar and bipolar arms trade systems, it will be useful to present some of his findings, although the extensive data available after he conducted his research will allow me to go much further in describing the postwar bipolar system.[7] Harkavy's diagrammatic model (figure 5–3) directly compares the interwar and 1945–65 systems.[8]

We can see an immediate change in figure 5–3 from the former to the latter period. The reduction of the system to two major powers who can produce the full line of military equipment, a system of alliances, and an international political system dominated by a zeitgeist of total ideological conflict has led to the outcomes depicted—predominantly single-client and within-bloc arms trade relationships, including coproduction and licensing agreements. Unlike the interwar system, arms trade patterns appear to correlate closely with the larger political system.

As for the larger international economic system, several linkages emerge. The within-bloc arms trade patterns can be partially explained by the two distinctly different economic systems, which inhibit cross-bloc transfers. Therefore, if such transfers occur, they will occur outside the dominant economic system (for example, grants, gifts, and aid). The natural resource and raw materials subsystem was basically controlled by the United States and the European colonial powers, meaning that new and developing states in the system had little leverage. As for military production, World War II had seen the largest

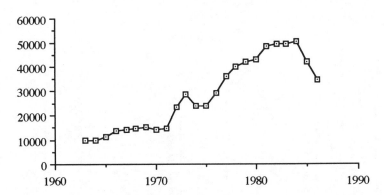

Figure 5–2. Arms Deliveries to the Third World, 1963–86

Source: Catrina, *Arms Transfers and Dependence*; from ACDA data.
Note: Figures in constant 1984 U.S. millions of dollars.

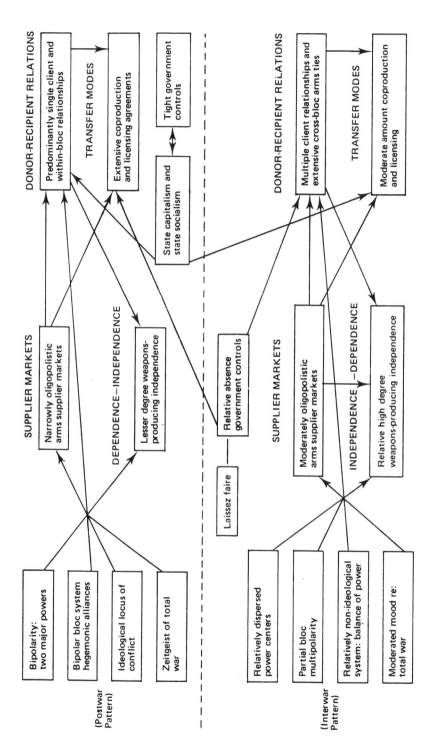

**Figure 5-3. Comparison of Postwar and Interwar Arms Trade Relationships**

Source: Harkavy, *The Arms Trade and International Systems*.

two suppliers totally committed to a major role for the state in this production. In the USSR, this had always been the case. In the United States, the military-industrial complex had become a reality, particularly with the advent of nuclear weapons and the perception that the nature of warfare had made the traditional mobilization approach obsolete. In such a system, private commercial arms trade simply disappeared as a phenomenon.

Although nuclear war, or its deterrence, was the major preoccupation of the two superpowers, regional, conventional armed conflict became the actual ground on which the cold war was fought. Despite both sides having nuclear weapons, proxy wars were fought with conventional weapons while the nuclear powers, eventually including China, France, and the United Kingdom, chose to use their weapons for deterrence only. This meant that despite a supposed revolution in warfare, little changed regarding the nature of actual armed conflict, and the arms trade tended to reflect this reality. It also should be added that the nuclear powers agreed not to transfer the capability that would create additional members of the nuclear club. This became clearer as the period wore on and the U.S. policy of massive retaliation gave way in the face of Dien Bien Phu, Hungary, Lebanon, and Vietnam. Both superpowers produced equipment for what they saw as the major challenge, a war in Europe. As a result, we saw the international military system and its dominant doctrines shape arms exports, as equipment received by superpower clients reflected the European requirements as opposed to any specific regional requirements.

The technology environment of this period is less predictive of actual arms trade patterns. It was a time of rapid rearmament and technological advances, particularly as the USSR began developing its nuclear arsenal in the late 1950s. However, two factors prevented the arms trade from being dominated by these technologies. First, both superpowers had tremendous surplus stocks remaining from World War II, and given the low technology base in emerging nations, these systems were quite adequate. The fact that they could be given as gifts with strings attached fit nicely with the bipolar alliance system. Secondly, these new technologies were relatively expensive, and the emerging nations (and those recovering in Europe) either had other uses for their capital or couldn't afford the newer equipment. Unlike the interwar years, when private arms manufacturers controlled relatively lower-technology armaments, the highly controlled arms trade of this period featured little trade in the more modern arms. The one exception was the U.S. effort through the Marshall Plan to get the defense industries of France and the United Kingdom back on their feet to participate in the rearming and defense of Europe. Harkavy's data on the modernity of equipment being traded clearly makes this point.[9]

## Characteristics of Units

The suppliers in this period were nation-states, mainly the United States, the USSR, and their respective allies. The United States used the Marshall Plan and other schemes to rebuild the arms industries of Western Europe, which were then able to supply their respective governments and, toward the end of the period, to become exporters. The industrial firms of the Western alliance were under close control of national governments, effectively reducing their role in a corporate sense to that of government organizations, especially when it came to exports. The USSR continued as a military economy, employing its satellites as arms producers and exporters. Private arms dealers, prominent actors in the previous system, disappeared except when used by either bloc to carry out covert arms exports at the behest of governments.

Table 5-1 summarizes the dependence of the nation-states of this system on external acquisition of major military equipment. Comparing it with a similar chart for the interwar period (see table 4-1), Harkavy concludes that the USSR, China, India, Canada, Sweden, and Australia were among those states increasing in independence (that is, able to produce more of their own defense equipment). Conversely, Czechoslovakia, Greece, the United Kingdom, Japan, Turkey, Denmark, Germany, Italy, Norway, Romania, and the Netherlands were states experiencing a decline in defense production capacity.[10]

The ratio of arms exports to overall production was fairly low, for several reasons. First, the two major suppliers had a significant amount of surplus equipment that was quite adequate for cold war purposes. The recipients/clients were involved in conflicts that required relatively unsophisticated equipment, needs met most often with surplus and older equipment. In the case of the United States, this is reflected in the preponderance of exports under the Military Assistance Program (MAP) (grants requiring no repayment by the recipients) as compared with either cash or credit transactions.

It should also be noted that the defense industrial firms in this system had more than enough to do either recovering from the war (for example, Western Europe) or supplying the domestic needs generated by the cold war. Defense budgets were large and backed by unprecedented public support, and there was no incentive or need to export much of the production. In 1966, the man appointed by the U.S. government to promote arms exports was still criticizing the "tendency of American companies to refrain from entering into the international arms market."[11] As for the USSR, their command and military-industrial economy eliminated the role of the firm by definition. In Western Europe, defense firms were recovering, but the overwhelming percentage of their sales were to their respective defense ministries to complete the post–World War II rearmament process. Beginning in the mid-1960s, coinciding with the U.S. sales drive to Western Europe, the United Kingdom and France began a major export drive of their own, basically putting to an end this period when NATO recipients either accepted grant aid or had to buy from the United States. The Marshall Plan had succeeded beyond anything imagined by the United States in the late 1940s.

## Table 5-1
### Summary of Postwar Dependence Levels by Weapons System

| Weapons System | Level | Countries |
|---|---|---|
| Aircraft | 1 | United States, USSR, France |
| | 2 | United Kingdom, Sweden |
| | 3 | West Germany, Japan, China, Italy, Australia, Czechoslovakia, Netherlands, Switzerland, Belgium |
| | 4 | India, Canada, Brazil, Spain, Argentina, South Africa |
| | 5 | Poland, Mexico, Indonesia, Austria, Yugoslavia, Turkey, Chile, UAR, Isreal, Taiwan |
| | 6 | All remaining nations |
| Naval | 1 | United States, USSR, United Kingdom, France, Japan |
| | 2 | West Germany |
| | 3 | Italy, Poland, Australia, Netherlands, Denmark, Yugoslavia, Finland, Norway, Portugal |
| | 4 | China, India, Canada, Brazil, East Germany, Spain, Argentina |
| | 5 | Mexico, Belgium, Hungary, South Africa, Indonesia, Turkey, Bulgaria, Colombia, Chile, Thailand, South Korea, North Korea, South Vietnam, Burma, Ceylon |
| | 6 | All remaining nations |
| Armor | 1 | United States, USSR, West Germany, United Kingdom, France |
| | 2 | Sweden |
| | 3 | Japan, Italy, Switzerland, South Africa, Yugoslavia |
| | 4 | China, India, Poland, Australia, Czechoslovakia, Brazil, Argentina, Israel |
| | 5 | Canada, Belgium, Dominican Republic |
| | 6 | All remaining nations |
| Small arms | 1 | United Staes, USSR, West Germany, United Kingdom, France, Czechoslovakia, Belgium, Switzerland |
| | 2 | Italy, Israel |
| | 3 | Japan, China, India, Poland, Australia, East Germany, Sweden, Netherlands, Spain, Argentina, Hungary, South Africa, Indonesia, Denmark, Austria, Yugoslavia, Finland, Norway, Bulgaria, Iran, New Zealand |
| | 4 | Canada, Brazil, Mexico, Portugal |
| | 5 | Romania, Pakistan, Turkey, Greece, Nigeria, Chile, UAR, Peru, Thailand, Cuba, South Korea, Taiwan, Burma, Dominican Republic, El Salvador, Paraguay, Rhodesia |
| | 6 | All remaining nations |

Source: Harkavy, *The Arms Trade and International Systems*, p. 193.

Table 5-2 seems to indicate that wealth equates with political power. The two exceptions appear to be Japan and West Germany, whose arms trading was limited by self-imposed constitutional restrictions.

The recipients of this period were of several types. First, former powers (such as West Germany, East Germany, China, and Japan) were rebuilding their

## Table 5-2
## Arms Transfers and GNP of Major Industrial States, 1960-64
*(billions of dollars)*

| Country | Average Value of Exports of Major Weapons to Third World | GNP |
| --- | --- | --- |
| USSR | 1,163.0 | 378 |
| United States | 1,137.8 | 805 |
| United Kingdom | 449 | 121 |
| France | 207.8 | 122 |
| West Germany | 34 | 173 |
| Japan | 11.2 | 123 |

Source: U.S. Arms Control and Disarmament Agency, *World Military Expenditures and Arms Transfers 1961-1971.*

forces. From 1950 to 1965, a significant percentage of the arms trade was to the industrialized world. For example, of more than five thousand jet combat aircraft exported by the USSR from 1949 to 1967, 55 percent went to industrialized countries. For the United States, 63 percent of its more than 5,200 jet aircraft exports went to industrialized countries.[12] From 1950 to 1965, NATO countries received a dollar value of $62.5 billion (in constant 1980 dollars), representing 57 percent of U.S. arms transfers to the entire world.[13] The data in figure 5-4 indicate how the ratio of industrialized to less-developed-country (LDC) recipients began to evolve during the 1960s, with a clear shift away from industrialized countries between 1964 and 1966.

Secondly, many recipients (such as Indonesia and Egypt) were clients of the two cold war protagonists and made significant use of the arms transfer instrument to gain influence or to enhance their security. If the U.S. response to the cold war was the establishing and arming of alliances ringing the USSR, the Soviet response was to jump the ring in 1955 and begin arming a new category of client (for example, Egypt, Syria, and Iraq).[14]

A third type of recipient developed as a result of the ending of the colonial era: newly independent countries establishing a military force for the first time. Of the forty-six countries with greater than 500,000 population who gained their independence between 1957 and 1981, thirty-nine did so prior to 1968.[15] A fourth type of recipient consisted of countries that were involved in regional conflict or had a central role in the playing out of the cold war. These countries needed military equipment more for military purposes than they did for prestige or as a mark of sovereignty. Listed in table 5-3 are the leading twenty-seven countries in terms of arms deliveries for the period 1961-65, representing 77 percent of all arms delivered during that period.

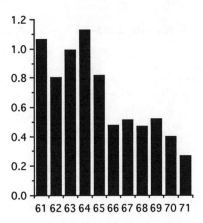

**Figure 5-4. Arms Imports Ratios of Developed versus LDC Recipients, 1961-71**

Source: U.S. Arms Control and Disarmament Agency, *World Military Expenditures and Arms Transfers 1961-71.*

## Structure and Stratification

**Suppliers.** Despite the fact that the United States and the USSR dominated this arms trade system, there was a steady increase in the number of suppliers. Harkavy found that for tanks and warships, the number of suppliers had increased since the interwar period from nine to eighteen for tanks and from nine to twenty-six for warships. This is misleading in that very few of these suppliers were independent enough from the superpowers to represent much of a real choice for the recipients. This can be seen in combat aircraft, where the number of suppliers actually decreased to sixteen from eighteen in the interwar period. The level of concentration (percentage of exports of the leading two suppliers) increased from 54 percent to 66.5 percent in tanks, and from 40 percent to 57.9 percent in combat aircraft.[16] Data compiled by ACDA shows that, between 1963 and 1966, the annual average of countries delivering arms was twenty-eight.[17]

**Recipients.** As for the number of recipients, there was real change here. The number of Third World nations importing arms rose steadily from twenty-five in 1951 to forty-seven in 1966 as decolonization became a reality.[18] Data compiled by ACDA shows that, from 1963 to 1966, the average number of nations (developed and less developed) importing arms in amounts greater than $2 million was seventy-eight.[19]

86 • *The International Arms Trade*

**Table 5-3**
**Leading Arms-Importing Countries, 1961-65**
*(billions of dollars)*

| Country | Arms Imports | Country | Arms Imports |
| --- | --- | --- | --- |
| West Germany | 1,779 | Iraq | 353 |
| Indonesia | 1,144 | Hungary | 332 |
| East Germany | 678 | North Vietnam | 316 |
| USSR | 675 | Taiwan | 293 |
| Egypt | 634 | Romania | 276 |
| Cuba | 622 | Greece | 256 |
| South Vietnam | 605 | United Kingdom | 218 |
| Poland | 593 | Australia | 217 |
| Italy | 534 | Pakistan | 214 |
| India | 531 | Japan | 209 |
| Czechoslovakia | 505 | Israel | 169 |
| South Korea | 417 | North Korea | 163 |
| Turkey | 397 | Iran | 119 |
| United States | 362 | | |

Source: U.S. Arms Control and Disarmament Agency, *World Military Expenditures and Arms Transfers 1961-1971*, p. A7.

**Market Shares.** Using the MIT data (the ACDA and SIPRI data were not yet available for his research), Harkavy compiled market-share data (table 5-4) for tanks, warships, and combat aircraft. He found a high correlation between this period and the interwar years for tanks and combat aircraft, whereas the market shares in warships had shifted drastically. (See table 4-3 for data on interwar years.)

In 1974, the first ACDA data on the arms trade was published (the initial data covered the 1961-71 time frame). Table 5-5 contains data on the leading suppliers for the period 1961-65, representing 96 percent of the arms trade of that period. SIPRI data on arms trade with the Third World during this period (table 5-6) reveals a similar picture.

**Supplier-Recipient Dependencies.** The data in tables 5-4, 5-5, and 5-6 picture an arms trade system dominated by the two superpowers as suppliers. Depicted in figure 5-5 are data on the relationship between suppliers and recipients regarding sole- versus multiple-supplier relationships. Harkavy's data show the patterns up to 1968, using mainly the MIT data base. When totaled, there are nineteen sole-supplier relationships, fifty-seven predominant-supplier relationships, and thirty-one rated as multiple supplier.[20]

SIPRI's data on arms transfers to the Third World figure 5-6 reveals that this period was dominated by sole and predominant relationships.

**Regional Distribution.** Figure 5-7 shows the regional distribution of the arms trade during the 1951-65 period. Note the importance of the Far East

Table 5-4
Postwar Market Shares for Tanks, Warships, and Combat Aircraft

| | Tanks | | | | Warships | | |
|---|---|---|---|---|---|---|---|
| Country | Rank | Number | Percentage of Total | Country | Rank | Number | Percentage of Total |
| USSR | 1 | 5,393 | 37.4 | United States | 1 | 116 | 27.8 |
| United States | 2 | 4,206 | 29.1 | United Kingdom | 2 | 87 | 20.9 |
| United Kingdom | 3 | 1,435 | 9.9 | Canada | 3 | 27 | 6.5 |
| France | 4 | 1,139 | 7.9 | Argentina | 4 | 18 | 4.3 |
| Egypt | 5 | 718 | 5.0 | Netherlands | 5 | 16 | 3.8 |
| Czechoslovakia | 6 | 250 | 1.7 | Brazil | 6 | 15 | 3.6 |
| West Germany | 7 | 250 | 1.7 | USSR | 7 | 15 | 3.6 |

Combat Aircraft

| Country | Rank | Number | Percentage of Total |
|---|---|---|---|
| United States | 1 | 2,512 | 29.7 |
| USSR | 2 | 2,390 | 28.2 |
| United Kingdom | 3 | 1,234 | 14.6 |
| France | 4 | 651 | 7.7 |
| China | 5 | 481 | 5.7 |
| Egypt | 6 | 352 | 4.2 |
| Sweden | 7 | 203 | 2.4 |
| West Germany | 8 | 201 | 2.4 |

Source: Harkavy, *The Arms Trade and International Systems*, pp. 60–78.

Table 5-5
Market Shares of Leading Exporting Countries, 1961-71

| Country | Arms Exports to All Countries | Percentage of World Total |
|---|---|---|
| United States | 6,002 | 36.7 |
| USSR | 6,170 | 37.8 |
| United Kingdom | 867 | 5.3 |
| France | 613 | 3.8 |
| Poland | 455 | 2.8 |
| Czechoslovakia | 459 | 2.8 |
| West Germany | 436 | 2.7 |
| Turkey | 269 | 1.7 |
| Canada | 261 | 1.7 |
| China | 106 | 0.6 |

U.S. Arms Control and Disarmament Agency, *World Military Expenditures and Arms Transfers 1961-1971*.
Note: Export figures represent ACDA figures for deliveries.

Table 5-6
Shifts in Supplier Market Shares, 1950-54 to 1960-64

| | 1950-1954 | |
|---|---|---|
| Country | Annual Average Exports of Major Weapons to Third World (millions of dollars) | Percentage of World Total |
| United States | 165 | 47.3 |
| USSR | 38 | 10.9 |
| United Kingdom | 76 | 21.8 |
| France | 17 | 4.9 |

| | 1955-1959 | |
|---|---|---|
| Country | Annual Average Exports of Major Weapons to Third World (millions of dollars) | Percentage of World Total |
| United States | 342 | 40.0 |
| USSR | 98 | 11.5 |
| United Kingdom | 171 | 20.0 |
| France | 71 | 8.3 |

| | 1960-1964 | |
|---|---|---|
| Country | Annual Average Exports of Major Weapons to Third World (millions of dollars) | Percentage of World Total |
| United States | 328 | 38.0 |
| USSR | 268 | 31.0 |
| United Kingdom | 111 | 12.8 |
| France | 63 | 7.3 |

Source: Brzoska and Ohlson, *Arms Transfers to the Third World: 1971-85*.

| Supplier | United States | United Kingdom | France | USSR |
|---|---|---|---|---|
| Sole | Bolivia<br>Nicaragua<br>South Korea<br>Liberia | Malawi<br>Gambia<br>Abu Dhabi<br>Bahrein<br>Muscat and Oman | Togo<br>Upper Volta<br>Senegal<br>Gabon<br>Chad<br>Central African Republic | Afghan<br>Guinea<br>Yemen<br>China |
| Predominant<br>(within-bloc pattern) | Colombia<br>Argentina<br>Brazil<br>Chile<br>Peru<br>Dominican Republic<br>Mexico<br>Rwanda<br>Thailand<br>Philippines<br>South Vietnam<br>All NATO<br>Japan<br>New Zealand<br>Australia<br>Spain | Rhodesia<br>Zambia<br>Kenya<br>Kuwait | Dahomey<br>Cameroon<br>Niger<br>Malagasy | North Korea<br>North Vietnam<br>Mali<br>All Warsaw Pact |
| Predominant<br>(cross-bloc pattern) | Iran<br>Ethiopia | Ceylon | None | Cuba<br>Iraq<br>Syria<br>UAR<br>Algeria<br>Cyprus<br>Somalia |
| | Multiple Supplier<br>(within West bloc) | Multiple Supplier<br>(within Soviet bloc) | Multiple Supplier<br>(cross-bloc) | |
| | Venezuela<br>Israel<br>Jordan<br>Lebanon<br>Saudi Arabia<br>Tunisia<br>Zaire<br>South Africa<br>Ivory Coast<br>Singapore<br>Sweden<br>Switzerland | none | Pakistan<br>Indonesia<br>Cambodia<br>Morocco<br>Mauritania<br>Tanzania<br>Yugoslavia<br>Nigeria<br>Austria<br>Ghana | Burma<br>India<br>Laos<br>South Yemen<br>Libya<br>Congo<br>Uganda<br>Sudan<br>Finland |

Figure 5–5. Postwar Supplier-Recipient Patterns, to 1968

Source: Harkavy, *The Arms Trade and International Systems*, p. 116.

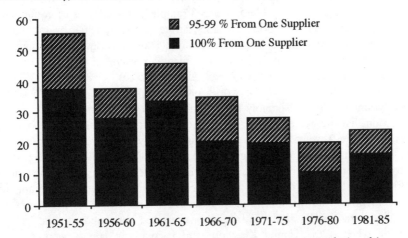

Figure 5–6. Sole- and Predominant Supplier-Recipient Relationships, 1951–85

Source: Brzoska and Ohlson, *Arms Transfers to the Third World: 1971–85.*

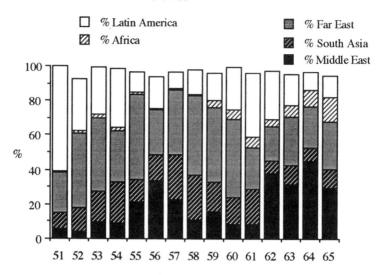

Figure 5–7. Regional Distribution of Deliveries of Major Weapons Systems to the Developing World, 1951–65

Source: Brzoska and Ohlson, *Arms Transfers to the Third World: 1971-85*.

throughout the period, with the Middle East rising significantly as the 1960s begin.

**Proliferation of Military Potential.** The changed structure is also reflected in the type of equipment that was being transferred, which was rarely top-of-the-line modern equipment. Table 5–7 lists a sample of those items of equipment traded to major Third World recipients during one year, 1964, by the major suppliers. Note that the United Kingdom and France, as new arms suppliers, were supplying comparatively newer equipment, a portend of things to come in the new arms transfer system that evolved in the 1970s.

What of the industrialized world, particularly those NATO and Warsaw Pact countries that presumably needed more modern equipment? There is no question that the United States sent its most modern equipment to Europe, at least until the Western European defense industries had recovered. A brief look at NATO inventories in this period reveals a trend toward more indigenous inventories in several countries, creating a different problem for an integrated alliance. By 1970, NATO inventories represented a mix between older grant aid equipment and newer indigenously produced equipment. The inventories of smaller states such as Italy and Belgium remained dependent on the United States, whereas the United Kingdom and particularly France were arming themselves with indigenously produced equipment. West Germany remained tied to U.S. equipment due to the West European Union (WEU) and self-imposed restrictions on defense production.

## Table 5-7
### Examples of Types of Weapons Systems Traded in 1964

| Supplier | Recipient | Equipment | Year of Production of First Model |
|---|---|---|---|
| USSR | Indonesia | MIG-21 fighter | 1956 |
| | Indonesia | AN-12 transport aircraft | 1958 |
| | Indonesia | Atoll AAM | 1956 |
| | India | PT-76 tank | 1955 |
| | Egypt | SA-2 SAM | 1957 |
| | Cuba | STYX antiship missiles | 1950s |
| United States | Taiwan | F-104G fighter | 1956 |
| | Pakistan | Tench class submarine | 1945 |
| | Pakistan | Sidewinder AAM | 1954 |
| | Greece | F-84F | 1946 |
| | Turkey | C-130E transport aircraft | 1952 |
| United Kingdom | Iraq | Hawker hunter-fighter | 1959 |
| | Jordan | Saladin APCs | 1958 |
| France | Israel | Mirage III-CJ fighter | 1962 |
| | Israel | SS-11 ATM | 1962 |

Source: Stockholm International Peace Research Institute, *The Arms Trade Registers*. Cambridge: The MIT Press, 1975.

## Modes of Interaction

**Modes of Payment.** The first point to make regarding the predominant mode of payment during this period was that it was government-to-government, a clear change from the interwar years. Defense industries made contracts with governments, who had almost total control of arms exports. Strictly commercial sales were minimal. For example, for the fiscal years 1950–63, U.S. commercial sales deliveries totaled $764 million, 2.3 percent of a total security assistance deliveries total of $32,762 million.

As previously discussed, surplus equipment from the stocks of the two superpowers dominated the arms trade in this period. As a result, the primary form of payment was grant aid. Readily available data on U.S. security assistance (figure 5–8) reveals that MAP grants dominated until the mid-1960s. It should be noted that beginning in 1966, the United States created a new MAP program for South Vietnam (known as MASF). If these are added to the picture, grants did not really decline until the early 1970s.

For most of this period, the recipients of arms lacked the cash for the outright purchase of equipment. As Western European nations began to recover, pressure mounted for them to pay cash or (at a minimum) obtain credit for such acquisitions. The U.S. mode of payment for the years 1963–66 is depicted in table 5–8. Note the rise of foreign military sales (FMS) deliveries, and the introduction of FMS and Export-Import (Ex-IM) Bank credits.

92 • *The International Arms Trade*

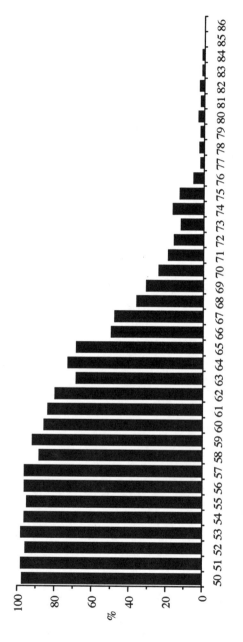

Figure 5-8. MAP Grants as a Percentage of Total U.S. Arms Deliveries

Source: Defense Security Assistance Agency, *Fiscal Year Series*, September 1987
Note: Excludes MASF funds to South Vietnam, 1966–75.

### Table 5-8
### U.S. Mode of Payment—FMS, Commercial, and MAP, 1963-66
*(millions of current dollars)*

| Mode of Payment | 1963 | 1964 | 1965 | 1966 |
|---|---|---|---|---|
| FMS orders | 1,401 | 1,261 | 1,579 | 980 |
| FMS deliveries | 480 | 703 | 822 | 1,052 |
| FMS credits | 75 | 110 | 317 | 323 |
| Ex-Im bank credits | 50 | 27 | 438 | 766 |
| Commercial sales | 208 | 155 | 196 | 237 |
| MAP appropriations | 1,004 | 957 | 975 | 877 |
| MAP deliveries | 1,256 | 1,435 | 1,062 | 942 |

Source: Defense Security Assistance Agency, *Fiscal Year Series*, September 1987. Washington: Department of Defense, 1987.

What of the Soviet Union, the other major arms supplier of this period? Prior to 1955, the USSR was busy shoring up socialist allies and buffer states (such as the Warsaw Pact nations, China, North Korea, and North Vietnam) with World War II surplus equipment. After 1955, it significantly shifted its foreign policy to include nonsocialist regimes as arms recipients in an attempt to increase global influence. Many of these recipients were poor, and it was at this point that the USSR began to use barter as a mode of payment. Many USSR arms recipients mortgaged their future by exchanging future production of commodities for used Soviet equipment, and the USSR maintained political leverage.[21]

As for that small segment of the arms trade conducted by such other Western suppliers as the United Kingdom, they, too, relied on surplus (especially the United Kingdom) until their industries had recovered. As this occurred, they began in the mid-1960s to contemplate a sales policy designed to distance them from the two superpowers. This was especially true of France and China, the latter becoming a major supplier after its break with the USSR in 1959, almost exclusively providing older equipment on a grant basis.

**Modes of Production.** Because grants dominated this system, the majority of recipients received equipment in the form of off-the-shelf deliveries, with little involvement in the production of the commodities. Many of the major arms deals of this period were accompanied by large numbers of military and technical advisors for long periods of time. Some of this was justified in terms of the inexperience of recipients, many of which were new countries. But these advisors also helped ensure that recipients toed the line in their use of the equipment to the benefit of the donor. The influence of these advisors was made all the more important by the fact that the cold war seriously reduced the free flow of technology in this system to within the two competing blocs. It was in 1949 that the Coordinating Committee (COCOM), a still-existing organization consisting of NATO plus Japan, was set up to stem the flow of technology to the communist states.

Within the two blocs, however, arms trade eventually evolved out of the grant aid stage to include a significant amount of licensing and some coproduction, mainly to industrialized recipients. For the Western alliance, this was part and parcel of the larger plan to develop integrated forces to deter a conventional attack from the east in Europe. Not surprisingly, a host of U.S. weapons systems were licensed for production in NATO countries. Japan developed a significant military inventory relying almost exclusively on licensed production by the end of this period. The USSR did the same with its allies, particularly China prior to 1959. Poland and Czechoslovakia produced submarines, tanks, surface ships, and fighter aircraft as part of the overall Soviet plan for military production and exports. Czechoslovakia was to play an important role as a surrogate arms exporter to the Third World for the USSR throughout this period. As for arms trade with the Third World, very few countries acquired arms via coproduction. One significant exception was India (MIG-21 fighters).[22]

*Regime Norms and Rules*

**Conflict.**

*Arms are exported as a function of regional and intrastate conflicts and military threats, which are almost always associated with the larger East-West conflict.* Unlike the interwar years, decisions to acquire and transfer arms in this system were significantly related to conflict. Major outbreaks of interstate violence in this period (Arab-Israeli in 1948 and 1956, India-Pakistan in 1965, North-South Vietnam in 1961–65, Korea in 1950–53, China-India in 1962) resulted in major acquisitions of arms by the participants. With the possible exception of India-Pakistan, these conflicts reflected and were fed by the larger East-West conflict. There were many civil wars and insurgencies in this period (for example, in Colombia, Cuba, Lebanon, Yemen, China, Indochina, Portugese colonies in Africa, and Algeria) that also sparked arms trade, although it was less linked to the larger global political conflict.

If conflict is defined as something more grand than a shooting war (for example, as the deterrence of rival blocs through the acquisition of military hardware), then we can see even more clearly how total the arms trade was related to conflict in this period. It was a period largely defined by military alliances and the arms flows intended to develop and maintain them. A brief return to the list of major recipients (table 5–3) confirms the absence of any recipient not connected in a major way to this reality. Newly independent countries were acquiring arms as a mark of sovereignty, but it was not a major feature of the system.

*Arms transfers are the major instrument of national security policy used within the two competing blocs to prepare for, deter, or wage armed conflict. Failure to use it will increase the chances for the outbreak of armed conflict.* What of the debate that raged in the interwar period regarding arms transfers causing or exacerbating conflict? The issue is almost totally absent, being replaced by a different focus. In the 1945–65 period, the dominant view of arms and conflict seemed to be that given the totality of conflict accompanying the cold war, arms transfers were the *only* instrument of foreign policy that could prevent conflict. Little debate existed in the Western democracies on this point, and Soviet (and later, Chinese) behavior confirmed its dominance as a guiding norm of the system.

**Foreign Policy and Diplomacy.**

*Arms transfers are the primary foreign policy tool used by the two blocs to signify and cement political alignment and bloc membership.* With the majority of the trade in this period being in the form of grant aid, combined with the pressure on the undecided to choose one or the other of the rival political blocs, arms transfers became the primary tool in the political competition. Grant aid was dictated by the fact that few of the pawns in the cold war had independent means of producing equipment for their own defense or for the national prestige owed a sovereign state. This meant that suppliers could and did attach significant strings to such aid, not to mention a large number of advisors and technicians to ensure compliance with these strings. It also became clear that much of the arms trade was not specifically related to military performance of the recipient. For example, Indonesia received massive quantities of military equipment almost totally unrelated to its military needs. The same could be said for U.S. arms to Pakistan; they were not enough to right an imbalance with India and seemed more related to ensuring its membership in the Central Treaty Organization (CENTO) and its overall political alignment with the West.

*Arms transfers are explicitly used to control and influence the behavior of recipients to support the foreign policy objectives of the supplier state. Trading arms for strategic access and arms embargoes are common foreign policy behavior.* Arms transfers were also used in a more specific way by suppliers to exert leverage and influence over recipients. Suppliers seriously reduced the foreign policy options of recipients by either withholding or restricting the use of arms supplies. All arms transfers were government-to-government, ensuring that these political objectives were explicit in any transfer decisions that were made at the national level.

**Economic Effects.**

*The government-to-government grant aid nature of the arms trade results in minimal economic costs or benefits to either supplier or recipient.* For most of this period, economics played a minor role in the pattern of the arms trade. First, the majority of the output of the world's defense industries went for domestic consumption. This was attributable to the large surplus remaining after World War II that could be exported, the inability of most potential recipients to absorb new weapon systems, and the fact that the United States and the USSR could achieve their foreign policy goals while using less than the most modern equipment. During this period, only a small percentage of the world's defense production was exported. For the defense industries, exports were not critical, so the concept of exports resulting in longer production runs and subsequently lower unit costs was not important in this period.

As for the recipient countries, their economies were either brand new or recovering from World War II and could not sustain any significant level of defense industrial production. This was less true for the recovering states in Western Europe, but as a rule, arms imports had little or no economic impact on recipient states.

**Arms Trade Control.**

*Significant control of arms exports exists at the unilateral/national level in terms of prohibiting private trade, limiting capabilities of exported weapons systems, and sending advisors and technicians to control their use.* There was a significant amount of arms trade control during this period, albeit at the national level. In the USSR, the command military economy, the totalitarian political system, and the single-minded anti-Western foreign policy ensured that every arms export was authorized at the highest level. The significant numbers of Soviet advisers accompanying these exports guaranteed that, once in the hands of the recipients, the arms would only be used to accomplish the goals of the supplier. Control was also exercised by not transferring those technologies and capabilities that would result in negative consequences for the supplier.

This national level of control also existed in the United States. Because grant aid (such as MAP) dominated arms exports, Congress had to appropriate funds on an annual basis, not only for equipment but also for the significant levels of advisory teams (such as Military Assistance Advisory Groups (MAAGs) sent along to control its use. The United States also minimized negative consequences by restricting the newer technologies and military capabilities to its reliable NATO partners and Japan. At the national level of analysis, it should be noted that an unusual level of political bipartisan support in the United States reinforced this control. It was also during this period that the United States began to focus on the Third World as the most critical arena in the struggle with the USSR. The Kennedy era saw the United States stressing counterinsurgency and, with it, a definite de-emphasis on those types of sophisticated weapons systems deemed inappropriate for this type of warfare. In 1965, the United States denied Peru the F-5A fighter on the grounds that Peru should not be spending its scarce resources on equipment that was ill suited for counterinsurgency. Ironically, this restraint or control was to open the Latin American market to the reinvigorated Western European suppliers and to increase significantly the levels of military spending in that region.

Can this control at the national level be accounted for by systemic factors? The cold war reached its peak during this period, during which it was perceived by all actors that regional conflict could quickly escalate to the superpower and nuclear level if it got out of control. This characteristic of the larger political system provided the incentive for the two major suppliers of the era to develop the implicit and unilateral arms trade control regime that evolved. It also dampened the appetite of the recipients. Another systemic factor that aided this control regime was the overall level of economic and technical development that existed in the majority of the recipients, especially those who might use their arms in ways detrimental to regional and international stability. These countries simply did not possess the wherewithal to develop military capabilities that could lead to an unstable conflict system, either regional or global. This is not to say that the superpowers always foresaw the negative consequences of their massive arms transfers. Rather, the trade was controlled in the sense that it was purposeful foreign policy.

*Attempts to control the arms trade at the multilateral/international level are nonexistent, except for within-bloc measures designed to enhance bloc capabilities.* What of international control of the arms trade? There were some examples of nations cooperating to stem the flow of weapons into troublesome arenas. For example, in May 1950 the United States, France, and the United Kingdom signed a tripartite declaration that created the Near East Arms Coordinating Committee. Since these three countries controlled all arms supplies to the Middle East, they were able to dampen arms acquisitions. When Egypt refused to swear allegiance to the Western powers in exchange for arms, the USSR stepped in with its major supply of arms to Egypt, effectively ending arms trade control in the region.[23] But overall, there was a marked lack of attention paid to those types of international control regimes that dominated the diplomacy of the interwar years. The first formal attempt to address control of the arms trade at the international level did not occur until 1965, when Malta put forth a proposal to publicize arms transfers through arms trade registers.[24] When viewed at the systemic level, this period amounted to a massive rearming of those states whose military capabilities were seriously diminished during the war, and arming of newly created states. World War II was *not* being explained in terms of arms races and the irresponsible behavior of private or governmental arms exporters. There was a systemic bias in favor of arms exports, as long as

98 • *The International Arms Trade*

they contributed to a bloc capability that either deterred war or allowed the bloc to prevail in the event of conflict.

Table 5–9 presents a summary of the arms trade system for the first twenty years of the postwar period.

**Table 5–9**
**Summary of 1946–66 Arms Trade System**

| | |
|---|---|
| *Boundaries, Linkages, and Environment* | • 1946–66<br>• Arms trade linked with bipolar political and ideological system<br>• Narrow oligopolistic arms-supplier markets are not linked with dominant laissez-faire economic system<br>• Advances in military technology have little impact on an arms trade dominated by older, surplus weapons systems |
| *Characteristics of Units* | • Suppliers are national governments<br>• Ratio of exports to production is low<br>• Recipients are new states, potential clients of the two superpowers, and pre-WWII powers who are rebuilding |
| *Structure and Stratification* | • United States and USSR dominate market<br>• Sole and predominant supplier-recipient patterns |
| *Modes of Interaction* | • Grant aid is predominant mode of payment<br>• Dominant modes of production are off-the-shelf to Third World, licensing within blocs |
| *Regime Norms and Rules* | |
| Conflict | • Arms are exported as a function of regional conflicts and military threats, almost always associated with the larger East-West conflict.<br>• Arms transfers are the major instrument of national security policy used within the two competing blocs to prepare for, deter, or wage armed conflict. Failure to use them will increase the chances for the outbreak of armed conflict. |
| Foreign Policy and Diplomacy | • Arms transfers are the primary foreign policy tool used by the two blocs to signify and cement political alignment and bloc membership.<br>• Arms transfers are explicitly used to control and influence the behavior of recipients to support the foreign policy objectives of the supplier state. |
| Economic Effects | • The government-to-government grant aid nature of the arms trade results in minimal economic costs or benefits to either supplier or recipient. |
| Arms Trade Control | • Significant control of arms exports exists at the unilateral/national level, in terms of prohibiting private trade, limiting capabilities of exported weapons systems, and sending advisors and technicians to control their use.<br>• Attempts to control the arms trade at the multilateral/international level are nonexistent, except for within-bloc measures designed to enhance bloc capabilities. |

## The Expanding Arms Trade System, 1966–80

*Boundaries and Environment*

The explanation of the transformation of the interwar system into that of the cold war period is straightforward, coinciding with the explanation for World War II itself. Once total war was on the horizon, acquisition of military capability became less a matter of arms trade and more a question of indigenous production. Similarly, the massive destruction caused by the war explains most of the structure of the postwar arms trade system, such as only the United States and the USSR capable of being suppliers.[25] Likewise, the regime rules flowed naturally from the nature of the cold war, which commenced in earnest in 1946.

The transformation of the 1946–66 system into what followed is not only more complex but also more significant, since the latter period was devoid of any event such as a major global war to provide easy explanations. This and subsequent transformations cannot be explained in terms of cataclysmic events such as World War II, but only in terms of those critical systemic events and trends, not related to the arms trade itself, that came together to foster change in the arms trade system.

In terms of trends, several were coming to an end that would have a significant impact on the arms-trading system and justify the choice of 1966 as the beginning of this system. First, as previously indicated, the Marshall Plan had worked, the Western European countries had recovered economically, and they were now producing arms of significant quantity and quality. Second, most of the new states of the postwar era (thirty-nine of forty-six) had achieved independence by 1967, meaning that they were steadily becoming more capable of acquiring major weapons systems and becoming legitimate recipients in the arms-trading system.

The tight bipolar political and ideological international system was beginning to unravel by the mid-1960s. The communist world was split when the USSR and China parted company in 1959, creating a new rivalry that had a significant impact on the arms trade. China, thanks to generous military assistance and licenses from the USSR, had become a major producer of military equipment and, in the process, an additional arms supplier in the arms trade system. Additionally, the new ideological rivalry between the two sparked regional arms races in the Third World, adding to the arms races occurring as a result of the East-West competition. In the West, France had reasserted itself after recovering from World War II, dropping out of the military structure of NATO, and becoming an independent producer and exporter of military equipment.

These changes in the international political system were reinforced by developments in the nuclear weapons balance between the superpowers. By the mid-1960s, it had become clear to the actors in the system (except perhaps for the superpowers themselves) that a fairly stable nuclear standoff had occurred. Certainly by the time following the Yom Kippur war in 1973, nuclear weapons arsenals had ceased to be the major factor in regional conflict initiation, exacerbation, or resolution. Rather, the actors in the arms trade system felt relatively certain that the conflict would not escalate into World War III, and as a result, arms exports soared.[26]

This is not to say that the cold war had ended. It had merely shifted to the developing world from the European theater, the latter having been stabilized by the nuclear balance and an adequate conventional-force balance. This playing out of the cold war in regional conflicts was to continue through the 1970s, perhaps coming to a close with the Soviet invasion of Afghanistan toward the end of that decade.

Major changes in the international economic system also had an effect on the arms-trading system. In 1970, the world went off the gold standard, signifying the end of the dominant role played by the United States since the Bretton Woods agreements of 1945. One of the immediate effects was a trade imbalance experienced by the world's major arms supplier, the United States, creating significant pressure to export all types of commodities, including arms. In 1973, the OPEC nations shocked the system with a quantum leap in oil prices. As with these other systemic changes, the arms-trading system kept pace. Oil-rich countries became customers for almost unlimited amounts of the expensive, modern, and high-technology weapons systems found in the inventories of the supplier countries.

The military technology environment also shifted drastically. The cost per ton of tanks, when adjusted for inflation, showed no upward or downward trend from 1918 to 1960, but after that increased by a factor of two or three times.[27] The impact was no different in other sectors. As a result, research and development began to consume as much as 50 percent of the cost of weapons production. Electronics became a critical part of military technology. This not only increased the cost of weapons production but also meant that production expertise became even more concentrated in those advanced industrial countries with electronics industries (the United States, Soviet Union, France, West Germany, the United Kingdom, Italy, Japan, and Sweden).

As with the previous system, we see a rather strong linkage between the arms trade system and the larger political, economic, technological, and military dimensions of the global system. As for when this system ended, this subject will be addressed as the next system is described in chapter 6.

*Characteristics of Units*

There was a great deal of continuity from the postwar system in terms of the typical supplier in the system. The national governments of the leading suppliers dominated the decision-making process, purely private traders and producers being relegated to mainly covert, government-sponsored transactions. The bulk of the arms produced and exported in this period were by arms industries under the control of governments. Many of the supplier rationales remained the same in this period, as the cold war continued to ensure that political motives dominated the system. At the end of this period, the definitive work on the arms trade by Andrew Pierre, entitled *The Global Politics of Arms Sales,* described the arms trade as "foreign policy writ large."[28]

But the new environment of the system did produce some changes in the typical arms-supplying country. Arms exports as a percentage of overall arms production increased significantly, as a result of the economic pressures to export. These pressures came both from the increased availability of oil money in countries lacking high-technology weapons systems and from the increasing costs of domestic production, which required more exports to offset these costs. The fact that surplus weapons systems were no longer available for export meant that all exports had to come from new or recent production, at least when compared with the previous period. The overall result of the economic and technological changes in the system produced governmental suppliers much more motivated to export arms, especially for the economic benefits.

The arms trade remained the province of the industrialized world, even though by the end of this period Third World arms producers had begun to export their wares. (This development would prove to be a forerunner of change in the 1980s). However, arms production at the middle and low levels of technology in Third World countries resulted in a decrease in the average magnitude of their arms imports (see figure 5–9).

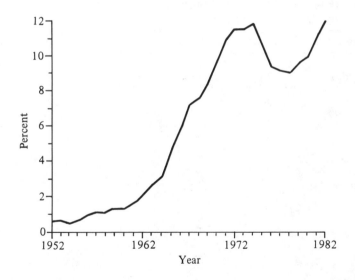

**Figure 5–9. Third World Major Weapon Production as a Share of Total Major Weapon Imports, 1952–82**

Source: Brzoska and Ohlson, *Arms Production in the Third World.*
Note: Five-year moving averages, in percentages.

Among the industrialized suppliers, a new type of supplier entered the system. the multinational corporation.[29] By 1975–76, Jane's was reporting on six multinational aircraft programs and five missile programs. Acquisition of these was restricted primarily to the countries that participated in the coproduction.[30]

During this period, the typical arms recipient was from the developing world and remained dependent on the industrialized world, particularly for advanced weapons systems containing high technology (such as electronics or engines). In 1967, 58.7 percent of the world's arms exports went to developing countries, with this figure rising to 75.5 percent by 1980 (an average of 67.4 percent for the fourteen-year period). Although the cold war continued through this period, the list of leading recipients in table 5–10 indicates that arms imports were becoming less political, at least as measured in East-West terms, and more related to regional disputes unconnected to the larger bipolar struggle. The international military environment, particularly the nuclear standoff resulting from mutual assured destruction, was having a definite effect on the proliferation of conventional military capability.

Table 5–10
**Distribution of Global Arms Imports**
*(percentages of arms delivered)*

| Country | 1963–67 | 1970–74 | 1976–80 |
|---|---|---|---|
| Vietnam | 10.1 | 5.5 | 2.7 |
| West Germany | 7.1 | 6.1 | — |
| USSR | 5.8 | 2.7 | 3.7 |
| India | 4.7 | — | 2.5 |
| East Germany | 4.2 | 3.9 | — |
| Poland | 3.8 | 3.1 | — |
| Turkey | 3.6 | — | — |
| Egypt | 3.5 | 6.8 | — |
| South Korea | 3.4 | 2.5 | 1.9 |
| Indonesia | 3.1 | — | — |
| Czechoslovakia | 3.0 | 2.5 | — |
| Italy | 2.5 | — | — |
| Israel | — | 4.8 | 3.8 |
| Iran | — | 6.2 | 8.0 |
| Iraq | — | 3.5 | 7.2 |
| Syria | — | 6.2 | 5.8 |
| Ethiopia | — | — | 2.1 |
| Algeria | — | — | 2.2 |
| Saudi Arabia | — | — | 4.8 |
| Libya | — | — | 7.7 |

Source: U.S. Arms Control and Disarmament Agency, various *World Military Expenditures and Arms Transfers* data.

Note: No entry indicates that country listed was not one of the top twelve importers for the period.

## Structure and Stratification

**Number of Suppliers.** The first structural change to be noted is that of the number of suppliers in the system. As the graphs in figure 5-10 depict, the number of states delivering arms (as defined by ACDA) almost doubles between 1971 and 1980, reaching a steady plateau of at least forty by 1978. A similar conclusion is reached if one looks only at major weapons systems transferred to the Third World (SIPRI data).

**Market Share.** Who were these new suppliers? They were mainly Third World and Western European states who were gaining the capability to produce military equipment based on licenses and (in some cases) indigenous capability. Figure 5-11 depicts how they began to export this production, with market share for "others" (Third World plus non-NATO European) rising from 3.8 percent to 7.7 percent, eventually peaking at 15.7 percent in 1984. Also note that the combined U.S. and USSR market share declined significantly, whereas NATO states increased their market share from 12.2 percent to 22.4 percent.

By 1980 a major inquiry had begun into the impact of these other suppliers on the system. It was common to find assessments that arms recipients were unleashing themselves from the superpowers, thanks to newfound suppliers with fewer strings attached to their armament exports. ACDA's 1980 essay on Third World arms production noted that "the proliferation of arms producers could well increase the difficulty of establishing multilateral controls over the international trade in conventional arms."[31] But the consensus at the end of this period of growth for Third World suppliers (in 1980) was that little impact had occurred. Most assessments predicted a tough road ahead, especially on the financial and technological dimensions. Neuman's seminal piece concluded that the addition of these so-called second-tier suppliers did not change the structure of the system. "The picture is not one of a zero-sum pie but of a growing pie

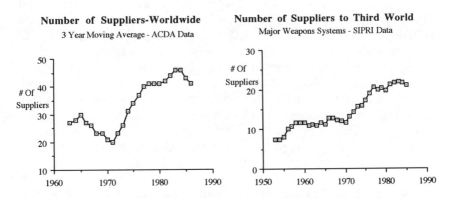

Figure 5-10. Number of Postwar Arms Suppliers

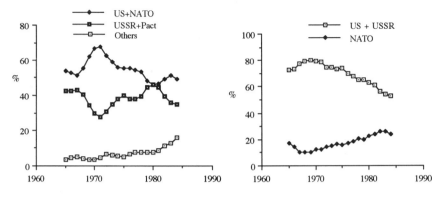

Figure 5-11. Market Shares of Global Arms Deliveries, 1965-84

Source: U.S. Arms Control and Disarmament Agency data.

with its proportions remaining unequal. Third world industries are advancing and integrating themselves into the established arms trade system rather than sustaining a challenge to it."[32]

This observation suggests that additional structural factors must account for changes in actual trading behavior. What becomes important is not only how many states supply arms but what type of arms are supplied. There is no doubt that the number of countries exporting major weapons systems increased between 1960 and 1980. Neuman's study concluded that in 1950, there were only five countries that could produce at least one major weapon system. That number grew to fourteen in 1960, twenty-one in 1970, and twenty-six by 1980.[33] The 1985 SIPRI assessment included the data shown in table 5-11.

These figures must be tempered by the fact that a significant percentage of the production was licensed, using imported technology. As of 1980, none of the systems in table 5-11 played a major role in the international marketplace. Of 638 arms deals involving Third World recipients listed in the 1981 SIPRI arms trade register, only 25 involved the systems and suppliers listed in the table.[34] In 1980, the value of production of major weapons systems by these second-tier suppliers was $1.45 billion, with $470 million being truly indigenous[35] and only a small percentage of that production being exported. During this time period, most of the exports from Third World producers were of low to mid-level technology (such as ammunition and small-to medium-caliber weapons systems) and had little impact on the overall trading patterns. The superpowers and Western Europe still produced and exported the advanced technology weapons systems.

**Supplier-Recipient Dependencies.** One implication of this increase in suppliers was that recipients now had much greater latitude choosing suppliers, with the potential for diversifying their dependence on commodities critical to their

## Table 5-11
## Third World Weapons System Production, 1966 versus 1980

| Weapons System | 1966 | | 1980 | |
|---|---|---|---|---|
| Fighters | India | | India | South Africa |
| | | | Brazil | Israel |
| | | | Taiwan | Argentina |
| | | | South Korea | |
| Helicopters | India | | India | Philippines |
| | | | Argentina | Indonesia |
| | | | South Korea | Brazil |
| | | | Egypt | |
| Missiles | India | Israel | India | Israel |
| | | | South Africa | Brazil |
| | | | Egypt | Taiwan |
| | | | Argentina | |
| Battle tanks | India | | India | North Korea |
| | | | Israel | Argentina |
| | | | Brazil | |
| Major fighting ships | North Korea | | North Korea | India |
| | | | Argentina | Brazil |
| | | | Peru | South Korea |

Source: Brzoska and Ohlson, *Arms Production in the Third World*, p. 23.

sovereignty and national defense. Neuman uses ACDA data on arms deliveries to conclude that "between 1972–73 and 1981–82, the number of Third World states receiving arms from four or more suppliers trebled, rising from 10 to 32. By 1981–82, 39% of the developing countries receiving arms were acquiring them from four or more sources, a rise of 24 percentage points over 1972–73."[36]

Excerpts from the SIPRI study (table 5–12) show this pattern of diversification. Although these data alone do not indicate actual diffusion of influence and dependency relationships, it would appear logical that increasing multiple-supplier relationships would eventually result in declining control by the superpowers via the instrument of arms transfers.[37]

**Number of Recipients.** One factor may help to explain why these additional suppliers did not have a greater impact on the system, and that is the number of recipients in the system. Figure 5–12 shows that between 1965 and 1980, thirty-three additional recipients of major weapons systems entered the market and logically consumed a significant portion of the wares offered by new suppliers. This is especially true given that most of these recipients were new countries. These second-tier suppliers were better equipped in both a technological and a political sense to deal with these countries.

**Regional Distribution.** Along with this proliferation of recipients came a shift in the regional distribution of arms transfers, away from Europe and toward those areas of the Third World that were experiencing conflict. Figure 5–13 depicts this shift. Soon after this period started, Europe began to be replaced as

Table 5-12
Number of Third World Countries Importing Major Weapons from More than One Supplier Grouping, 1971-85

| Joint Supplier Grouping | Number of Recipients Supplied | | | | | | | | | | | | | | |
|---|---|---|---|---|---|---|---|---|---|---|---|---|---|---|---|
| | 1971 | 1972 | 1973 | 1974 | 1975 | 1976 | 1977 | 1978 | 1979 | 1980 | 1981 | 1982 | 1983 | 1984 | 1985 |
| USSR + U.S. | 3 | 3 | 1 | 1 | 4 | 6 | 5 | 7 | 4 | 6 | 6 | 3 | 4 | 3 | 3 |
| U.S. + NATO-Europe | 24 | 20 | 21 | 23 | 26 | 25 | 27 | 30 | 21 | 29 | 32 | 29 | 31 | 26 | 18 |
| USSR + NATO-Europe | 5 | 4 | 4 | 3 | 5 | 11 | 10 | 13 | 14 | 12 | 11 | 7 | 9 | 7 | 9 |
| USSR + Third World | 0 | 0 | 2 | 2 | 1 | 3 | 3 | 3 | 9 | 5 | 5 | 3 | 3 | 4 | 2 |
| U.S. + Third World | 4 | 5 | 3 | 3 | 5 | 9 | 4 | 7 | 6 | 8 | 12 | 9 | 15 | 8 | 9 |
| NATO-Europe + Third World | 3 | 0 | 4 | 4 | 4 | 9 | 7 | 12 | 12 | 13 | 15 | 16 | 18 | 12 | 9 |

Source: Brzoka and Ohlson, *Arms Transfers to the Third World: 1971-85*.
Note: NATO-Europe refers to West European NATO member countries.

Arms Trade Systems in the Postwar Period • 107

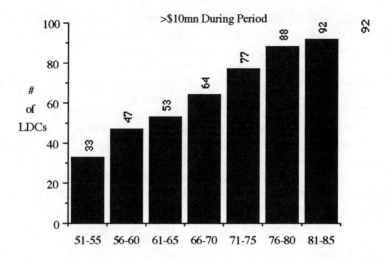

**Figure 5-12. Number of LDCs Receiving Major Weapons Systems, 1951-85**

Source: Brzoska and Ohlson, *Arms Transfers to the Third World: 1971-85*.
Note: LDCs receiving weapons systems of value greater than $10 million during period.

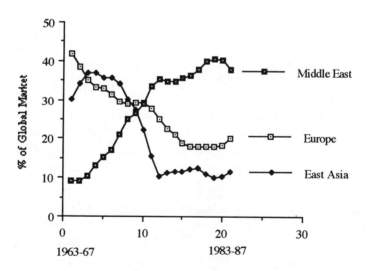

**Figure 5-13. Regional Distribution of Arms Deliveries, 1963-87**

Source: U.S. Arms Control and Disarmament Agency, various *World Military Expenditures and Arms Transfers* data.

108 • *The International Arms Trade*

the leading market, first by East Asia (the Vietnam war) and then by the Middle East. The lists in table 5–13 of the top dozen recipients during different periods reinforces this conclusion.

**Proliferation of Military Potential.** A final structural factor is the proliferation of military potential inherent in the increased trade. The number of aircraft, missiles, armored fighting vehicles, and warships transferred to the Third World from 1970 to 1976 equaled those transferred from 1950 to 1970. The average annual increase in such equipment was 5 percent annually from 1960 to 1966, rising to 15 percent annually from 1970 to 1976. The transfer of 18,607 tanks, 243 naval warships, 4,510 supersonic aircraft and 18,175 surface-to-air missiles between 1967 and 1976 represented a major influx of equipment for use in conventional warfare. Though 1960 saw MIG-19, F-86, and a few F-104 fighter aircraft transferred, by the mid-1970s the trade included the most capable aircraft in the superpower inventories: AWACS, F-16, and MIG-23. When regional conflicts did erupt (such as the 1973 Yom Kippur war), the damage and casualties imparted by the participants was in no small way related to this increased proliferation of capability. The number of countries possessing supersonic aircraft went from twelve in 1966 to fifty-three by 1980; for missiles, the increase was from twenty to fifty-three.[38] The sample in table 5–14 of weapons transferred to developing countries in 1976 provides a definite contrast to the similar list for 1964.

*Modes of Interaction*

**Modes of Payment.** The dominant mode of payment during this period continued to be the government-to-government transaction, but there was a

Table 5–13
Top Twelve Arms Recipients, 1963–67 to 1976–80

| 1963–67 | 1970–74 | 1976–80 |
|---|---|---|
| South Vietnam | Egypt | Iran |
| West Germany | Syria | Libya |
| USSR | Iran | Iraq |
| India | West Germany | Syria |
| East Germany | Vietnam | Saudi Arabia |
| Poland | Israel | Israel |
| Turkey | East Germany | USSR |
| Egypt | Iraq | Vietnam (Dem. Rep.) |
| South Korea | Poland | India |
| Indonesia | USSR | Algeria |
| Czechoslovakia | South Korea | Ethiopia |
| Italy | Czechoslovakia | South Korea |

Source: U.S. Arms Control and Disarmament Agency, various *World Military Expenditures and Arms Transfers* data.

## Table 5-14
### Examples of Types of Weapons Systems Traded in 1976

| Supplier | Recipient | System | Year of Initial Product |
|---|---|---|---|
| United States | Chile | Maverick AGM | 1972 |
| | Iran | F - 14A fighter | 1974 |
| | Iran | Phoenix AAM | 1971 |
| | Iran/South Korea | Harpoon SSM | 1975 |
| | Israel | F-16 fighter | 1976 |
| USSR | India | Nanunchka missile corvette | 1971 |
| | Iraq | MI-24 attack helicopter | 1973 |
| | Libya | TU-22 fighter | 1972 |
| France | Abu Dhabi | Crotale SAM | 1968 |
| | Abu Dhabi | Mirage 5 fighter | 1969 |
| | Argentina/Brazil | Exocet SSM | 1972 |
| | Egypt/Iraq/Kuwait | Mirage F1-E | 1976 |
| U.K./France | Oman | Jaguar fighter | 1972 |
| West Germany | India/Peru | Type 209 submarine | 1974 |

Source: *SIPRI Yearbook 1977.*

clear shift away from grants and aid to credit, and eventually to cash, when a surplus of petrodollars was created by the oil price increase in 1973. Data on the United States (see figure 5-14) is indicative of the system as a whole.

The behavior of the other major supplier in the system,[39] the USSR, also reflected this systemic shift. The data in table 5-15 on Soviet hard currency sales indicates a response to the sudden availability of petrodollars.[40] This trend is made even more clear when Soviet arms trade with its oil-rich partners Algeria, Iraq, and Libya are noted. These three countries imported $11.4 billion in arms from the USSR from 1975 to 1979; this represented 73 percent of all of their arms imports (see table 5-16). Further, exports to these three represented 45 percent of all Soviet arms exports. Although the USSR mode of payment continued to be predominantly aid, hard currency became a primary motive in almost half of their exports. The effect of this change in motive was to become particularly clear when these states began to behave in ways unlike the typical subservient Soviet client.

**Modes of Production.** In the interwar arms trade system, off-the-shelf trade in end items and licensed production were the dominant modes of production. In the immediate postwar period, the two superpowers dominated a trade almost exclusively in end items, with licensed production being restricted to cooperative programs in NATO and the Warsaw Pact countries. In the 1966–80 period, with the massive stocks of World War II surplus equipment no longer available for export, several new trends in production became common in the arms trade system. First, the quest for independence in national security affairs in most of the developing world resulted in the worldwide proliferation of

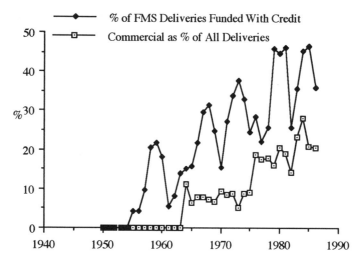

Figure 5-14. Growth of U.S. Commercial and Credit Exports

Source: Defense Security Assistance Agency, *Fiscal Year Series,* September 1987. Washington: Department of Defense, 1987.

arms deals featuring licensed production by the recipient. Figure 5-15 depicts this trend.[41]

The second change was the emergence of trade in advanced weapons systems produced by Western European multinational corporations. Because of the continued ability of the cold war to discipline the behavior of Western European countries, and the dominance of U.S. technology, these multinational firms were not truly independent in the sense that multinational firms such as AT&T were. At first, their production was consumed by those nations whose industries participated in the consortia, but eventually the multinational production was transferred throughout the system. One effect of this change was the decline in accountability when it came to the effects of the transfer. This tended to introduce the idea that arms could not lead to political influence in the traditional sense if the supplier was a multinational firm rather than a nation-state. As for any negative consequences, no single actor could be identified when it came time for remedies.

*Regime Norms and Rules*

**Conflict.**

*The conduct and outcome of interstate regional conflict is determined primarily by U.S. and Soviet arms transfer policy and behavior.* Among the

### Table 5-15
### Soviet Hard Currency Sales, 1970–1981
*(millions of constant 1972 U.S. dollars)*

| Year | Arms | Oil[a,b] | Gas[b] | Gold |
|---|---|---|---|---|
| 1970 | 437 | 423 | 13 | Negligible |
| 1971 | 417 | 591 | 21 | 25 |
| 1972 | 600 | 556 | 23 | 289 |
| 1973 | 1,513 | 1,180 | 22 | 910 |
| 1974 | 1,303 | 2,228 | 75 | 594 |
| 1975 | 1,192 | 2,525 | 175 | 576 |
| 1976 | 1,398 | 3,411 | 262 | 1,034 |
| 1977 | 2,299 | 3,779 | 404 | 1,155 |
| 1978 | 2,636 | 3,796 | 707 | 1,677 |
| 1979 | 2,359 | 5,863 | 859 | 912 |
| 1980 | 2,315 | 6,733 | 1,515 | 884 |
| 1981 | 2,148 | 6,285 | 2,023 | 1,381 |

Source: Calculated from data presented in Joan Parpart Zoeter, "U.S.S.R., Hard Currency Trade and Payments," in Joint Economic Committee. Congress of the United States, *Soviet Economy in the 1980s: Problems and Prospects. Part 2* (Washington, GPO, 1983), pp. 503–4.

[a] Connotes petroleum and petroleum products.
[b] Excludes exports for hard currency to other communist countries.

four categories of regime rules—conflict, foreign policy and diplomacy, economic effects, and arms trade control—the least amount of change occurred in the rules and norms that guided the use of arms transfers in relation to conflict. The locus of conflict shifted from Europe to the developing world, and within that world from East Asia to the Middle East. However, most of the conflicts requiring and influenced by major weapons acquisitions remained greatly influenced by the larger East-West conflict. The best example is the Arab-Israeli war in October 1973, in which all sides were armed exclusively by

### Table 5-16
### Soviet Arms Transfers to Major Clients, 1975–79
*(millions of dollars)*

| Country | Total Arms Imports | Arms Imports from USSR |
|---|---|---|
| Algeria | 1,900 | 1,500 |
| Cuba | 8,75 | 875 |
| Ethiopia | 1,800 | 1,500 |
| Iraq | 6,800 | 4,900 |
| Libya | 6,900 | 5,000 |
| Syria | 4,500 | 3,600 |
| Vietnam | 1,300 | 1,300 |
| Warsaw Pact | 7,875 | 6,950 |
| Algeria, Iraq, and Libya (combined) | 15,600 | 11,400 |
| Totals | 31,950 | 25,625 |

Source: U.S. Arms Control and Disarmament Agency, *World Military Expenditures and Arms Transfers 1970–79*.

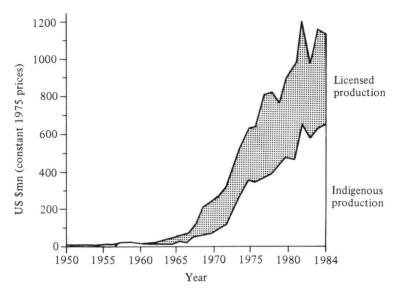

**Figure 5-15. Value of Production of Major Weapons in the Third World, 1950-84**

Source: Brzoska and Ohlson, *Arms Production in the Third World*, p. 9.

the superpowers. The war may have started as a result of indigenous causes, but once it was under way, the two arms suppliers had a significant impact on the outcome. In the case of the United States, Israel's strategic plans and operations were effectively shaped by the sole-supplier arms relationship with the United States.[42] Neuman's research on the impact of arms transfers on the wars that occurred late in this period (Somalia-Ethiopia, Western Sahara, China-Vietnam, the civil wars in Central America, Iran-Iraq, Afghanistan, Argentina-United Kingdom in the Falklands, Lebanon in 1982) makes a convincing argument that

> neither war nor the rising number of weapon suppliers have served to destabilize the international system or reduce the influence of the United States and USSR in it. Although diversification and indirect sources of supply have permitted Third World combatants to continue fighting, often at tremendous cost in lives, the type of weapons and training available to them has remained limited. In effect, the superpowers have retained control over the quality if not the quantity of the arms trade, and ultimately, therefore, over the level of sophistication at which wars can be fought in the Third World.[43]

Neuman's overall argument is that despite the fact that more suppliers and a greater variety of available equipment produced more choices for recipients engaged in conflict, the superpowers retained significant power to influence conflict outcomes for five basic reasons. These include superpower restraint and arms transfer restrictions; the lack of human resources in the recipient countries; economic restraints; the demand for fast, large deliveries; and the demand for intelligence information.[44]

*The control of the outbreak of interstate conflict by arms suppliers, especially the superpowers, is very difficult given the proliferation of military capability in the system.* The fact that the *outcome* of major interstate wars was determined by the superpowers did not mean either that their control was total or that it necessarily deterred the *outbreak* of the conflict. One result of the proliferation of military capability was to exacerbate tensions and, in some cases, to promote the outbreak of conflict. In the aforementioned 1973 Middle East war, Egypt asked all of its Soviet advisors to leave and in effect started the war on its own, even though it fought with equipment totally supplied from its superpower ally. The same could be said for Israel in both the 1967 and 1973 wars. Assuming that both the USSR and the United States had no interest in their clients actually starting wars, the evidence seems to show that the superpowers did not have the capability to use arms exports to control the outbreak of war. This rule certainly would apply to all of the Indian-Pakistani wars, as well as most others in this period. Perhaps the example of the U.S. attempts to deter Turkey from invading Cyprus is a good one. In the postwar system, the United States was able to deter such an attack in 1964 when President Johnson threatened Turkey with an arms embargo, among other things. In 1974, although a similar threat was implicitly made, it had no effect, as Turkey felt it had more leverage over the United States than it did ten years earlier. One of these sources of leverage was increased opportunities for resupply from countries other than the United States, a characteristic of the new arms trade system.[45]

*Intrastate conflict and terrorism are not affected by the arms-trading system, as systemwide proliferation of low- to mid-level weapons systems allows both government and oppositionist participants to use multiple sources for weapons, including the private commercial market.* Unlike interstate conflict, intrastate and internal conflict became fairly independent from the larger arms-trading system of this period. This was a change from the earlier postwar period, when a polarized political system saw the superpowers supply arms to combatants in even the smallest conflict. The low level of military technology required for this type of warfare, combined with the increasing availability of such weapons from suppliers who showed little incentive for control or restraint, meant that the commencement, exacerbation, conduct, and outcome of such wars had little to do with the type or source of weapons involved. One of the best examples from this period is the insurgency in Nicaragua that ended with the overthrow of the Somoza government in 1979. Israel supplied the Somoza government to the bitter end, despite pressure from the United States to stop. The Sandinistas, although clearly supported by the Soviet bloc and especially Cuba, had a variety of choices when it came to where it could obtain weapons systems, including the open market. Terrorism increased significantly in this period, one reason being the increased availability of arms on the open market. Despite the prevailing argument that the USSR was responsible for much of the world's terrorism in this period, arming the terrorist was not one of the requirements for the USSR.

*Third World recipients of modern weapons systems do not have the quantity, infrastructure, or manpower skills required to use these systems effectively in armed conflict or to threaten the major powers militarily.* There was a great deal of concern that the proliferation of high-technology weapons systems would create the potential for negative consequences once armed conflict ensued. There was no shortage of analysts who predicted a shift in military power to Third World recipients of these newer systems. In reality, there was little difference in the casualties incurred between both periods. Neuman concludes that, if anything, possession of these modern systems made the recipient military forces less effective, since most did not have the training or infrastructure to take proper advantage of the technology and, in the end, had to rely on their conservative and conflict-averse donors for assistance.[46] There were some exceptions, notably Israel. But in the main, recipients either remained under the control of their suppliers (USSR-Syria) or simply did not use the equipment and reverted to using mid- to low-level weapons systems (e.g., Iran-Iraq during the first years of their war). A related concern that also did not materialize was the fear that these Third World countries could threaten the supplier countries with the systems.

*The presence and sudden availability of first-line military equipment, attributable to systemic factors of bloc obsolescence and rising oil prices, creates significant demand in the developing countries for these weapons systems for prestige purposes. The threat of armed conflict is no longer a prerequisite for receiving or exporting modern military equipment.* Unlike previous periods, arms acquisitions no longer had to be justified based on the threat of immediate conflict. Once the petrodollar influx had occurred, and the major suppliers began to experience serious budgetary pressure to export their production, arms exports for reasons of boosting the recipient's prestige, filling power vacuums, or creating "stability" became an accepted rationale for both supplier and recipient. The top four recipients of arms deliveries in this period—Iran, Libya, Iraq, and Saudi Arabia—were not considered frontline states in the Arab-Israeli conflict. Despite some interstate tension and hostility (such as between Iran and Iraq), the suppliers rarely were concerned about the impact their exports would have on either the commencement, exacerbation, or conduct of armed conflict.

**Foreign Policy and Diplomacy.**

*Using arms transfers as a bargaining tool to gain political influence continues to dominate as a rationale for the major suppliers, particularly in regard to gaining strategic access in the Third World.* At the nation-state level of analysis, political influence continued to remain as a primary motive for the export of arms. The United States and the USSR continued to supply states with arms as they attempted to influence the short-term behavior of clients in the larger East-West contest. Certainly the massive influx of arms sent into Angola, Mozambique, and Ethiopia by the USSR is explained primarily by this motive. A brief review of U.S. State Department justifications for the major sales of this period reveals similar motives on the part of the United States. Pierre's 1982 book summing up this period, *The Global Politics of Arms Sales,* is correct in highlighting the political nature of a significant percentage of the arms trade, especially by the superpowers.[47] Even France's motives for its increased trade had a vital political component. A corollary to this rule is that arms are used extensively to gain access to overseas bases and other forms of strategic access. For example, the Soviet navy becomes a blue-water navy during this period, and much of the Soviet trade was related to this development. Third World recipients had not yet gained the political clout to resist this effort, although as the bipolar system loosened this began to change.

*Arms transfers are not a reliable bargaining tool for influencing the short-term behavior of clients in a system where recipients are acquiring arms less for military reasons and more for prestige and regional political power, especially when they can spread their dependence among several suppliers, develop indigenous systems, or accomplish national security objectives with less modern equipment.* The proliferation of suppliers in the system, a loosening of the bipolar political system, a decline in the role of ideology, and the nuclear parity or standoff between the superpowers resulted in a major change in the relationship between arms and influence. Although the major powers continued to act with influence as a primary rationale, the actual utility of this instrument when viewed as a bargaining tool produced a different reality. The increase in multiple-supplier relationships caused a natural blurring of commitment between supplier and recipient. Recipients increasingly parlayed this multiple-supplier reality into a decrease in dependence, at least on any one supplier. This was especially true in those cases where recipients were not faced with immediate armed conflict, which (as pointed out above) would have increased the potential for influence on the part of the supplier. In the end, suppliers could not prevent major defections by major client states, the ultimate test of influence. Table 5–17 is a brief summary of some of the supplier-recipient shifts that occurred in this period.[48]

This situation was further exacerbated by the presence of multinational suppliers in the system. Although these corporations started out to supply just

**Table 5-17**
**Examples of Bloc Changes in Arms Supply, 1968-77**

| Country | Supplier New | Supplier Old | Approximate Date of Change | Reason |
|---|---|---|---|---|
| Cambodia | China | U.S. | 1973 | Regime change |
| Laos | USSR | U.S. | 1973 | Regime change |
| Ethiopia | USSR | U.S. | 1977 | Regime change |
| Somalia | West | USSR | 1977 | Change in Soviet policy |
| Egypt | West | USSR | 1974 | Decreased dependence |
| Sudan | West | USSR | 1974 | Decreased dependence |
| Iraq | Cross | USSR | 1975 | Decreased dependence |
| Peru | Cross | West | 1973 | Decreased dependence |
| Libya | Cross | West | 1970 | Oil money |
| North Yemen | West | USSR | 1974 | Change in policy |
| Zambia | Cross | West | 1971 | Nonaligned policy |
| Congo | East | France | 1968 | Shift in regime's policy |

Source: Mihalka, in Neuman and Harkavy, *Arms Transfers in the Modern World,* table 4-10.

those industrialized states who participated in the actual coproduction of the systems, as Hagelin's data (figure 5-16) show, the exports to the Third World of these systems were not insignificant.[49]

In the case of pariah states, their regimes could now stay in power longer because of the increased availability of those mid- to low-level weapons systems needed to maintain internal stability. All of this is not to say that arms transfers did not lead to influence. A significant amount of influence *was* obtained using this instrument, not only by the superpowers but by Western European states (such as France when dealing with its ex-colonies). The point to be made is that unlike the postwar system, political influence viewed as bargaining was a lot harder to come by.

*Arms transfers result in structural influence of the patron over the client's medium-term behavior and ensure stability in the dominant supplier-recipient relationships.* Krause's recent work on influence suggests that when influence is defined as structural, as opposed to bargaining, a great deal more of it is to be gained when using arms transfers. Krause states that "structural power is exercised when a patron alters the range of options open to the client or makes it more or less costly for the client to change these options."[50] An example from his work on the Middle East is the U.S. relationship with Egypt, Israel, and Jordan in the 1970s and early 1980s:

> Although [the U.S.] ability to manipulate arms transfers for specific gains in the peace process was limited, the general goal of avoiding the outbreak of a war was assured by the balancing of arms supplies to both sides. This did not involve serious restrictions on the military options that Egypt and Israel could pursue (although it did for Jordan), but its dominant position means the United

| Project | Year of agreement | Partners | Project shares (per cent) | Joint managing company | Numbers ordered by 1983 from | | | Third World share of orders (per cent) |
|---|---|---|---|---|---|---|---|---|
| | | | | | Third World countries | Other industrialized countries | Producer countries | |
| *Aircraft* | | | | | | | | |
| Jaguar | 1965 | UK (British Aerospace)<br>France (Dassault) | 50<br>50 | Sepecat, UK | 175 | 0 | 402 | 30 |
| Alpha Jet | 1969 | France (Dassault)<br>FR Germany (Dornier) | 50<br>50 | | 123 | 33 | 350 | 24 |
| Tornado | 1968 | UK (British Aerospace)<br>FR Germany (MBB)<br>Italy (Aeritalia) | 42.5<br>42.5<br>15.0 | Panavia, FR Germany | 0 | 0 | 809 | 0 |
| *Helicopters* | | | | | | | | |
| SA-330 Puma/<br>SA-332 Super Puma | 1967 | France (Aérospatiale)<br>UK (Westland) | 72.5<br>27.5 | Heli-Europe Industries, UK | 394 | 49 | 204 | 61 |
| SA-341/342 Gazelle | 1967 | France (Aérospatiale)<br>UK (Westland) | 65<br>35 | | 244 | 255 | 571 | 23 |
| WG-13 Lynx | 1967 | UK (Westland)<br>France (Aérospatiale) | 70<br>30 | | 50 | 62 | 234 | 14 |
| *Missiles* | | | | | | | | |
| Milan | 1965 | France (Aérospatiale)<br>FR Germany (MBB) | 50<br>50 | Euromissile, France | Total orders: 300 000[a] | | | [54[b]] |
| HOT | 1965 | France (Aérospatiale)<br>FR Germany (MBB) | 50<br>50 | Euromissile, France | Total orders: 50 000[a] | | | [64[b]] |
| Roland | 1966 | France (Aérospatiale)<br>FR Germany (MBB) | 50<br>50 | Euromissile, France | Total orders: 25 000[a] | | | [5[b]] |

[a] According to Euromissile.
[b] Share of actual exports to all countries in relation to procurement by France and FR Germany up to the end of 1982.

**Figure 5–16. Selected Multinational Weapons Projects and Export Sales, 1965–83**

Source: *SIPRI Yearbook 1984*, p. 154.

States has sufficient structural power to make defection from the "no war" regime unattractive. By contrast, the Soviet Union has had virtually no structural power to promote its broad geopolitical aims in the region since 1976.[51]

Much the same point could be made about the USSR and India, France and its ex-colonies, and perhaps West Germany and those states that acquired its expensive high-technology submarines. Recipient states were free to take advantage of the loosening of the international system and the emergence of more suppliers. That many did not is attributable to the high costs of defection associated with structural influence.

**Economic Effects.**

*Major arms-producing states must either export a significant percentage of their production of advanced systems or expire as independent actors.* Some of the most significant systemic changes occurred on the economic dimension. The sudden availability of petrodollars to the OPEC states and those to whom they loaned money, combined with the rising cost of the advanced weapons systems needed by the superpowers and their allies in the cold war, resulted in economic motives becoming critical elements in the decision-making process of both suppliers and recipients. For the typical major supplier in this system, defense budgets came under significant pressure. It was faced with a choice of discontinuing the production of certain weapons systems, a declining inventory of these advanced higher-cost items, or the acquisition of a system from another state. Any of these outcomes would have resulted in a decrease in independence and power. To minimize this happening, it chose the option of maximizing the export of its production. Even the USSR, with its military command economy, began to behave according to this new norm.

*Economic pressures make major suppliers, and particularly their arms industries, more likely to export technology and other aspects of the production process in order to survive as independent producers of advanced equipment.* This pressure to "export or expire" also meant that suppliers were under increased pressure from recipient states, especially the ones without access to petrodollars, to part with those elements of the production process that might result in improving the military-industrial capability of the recipients. The general looseness of the arms trade regime, spurred on in no small part by the détente between the superpowers and a more open trade system in general, meant that hard-pressed defense industries were under much more pressure to export whatever it took to survive.

*The sudden acquisition of convertible cash pressures many recipient states and their allies to recycle petrodollars by acquiring very advanced and expensive weapons systems, exacerbated by the "prestige race" that ensues. Poorer states with greater security needs tend to be less preferred by suppliers needing the cash.* From the recipient perspective, the systemic changes meant that, all of a sudden, the most advanced weapons systems could be acquired for the asking. Since most of the states in this category were interested in arms for prestige, pressure mounted to keep up with other states. Further, since cash sales became the normal mode of payment in the system, traditional and poorer allies of the major powers had significantly less bargaining power when it came to acquiring weapons systems. Additionally, for most of these recipients, there was no way that they could actually use their newfound wealth for domestic development. The acquisition of expensive weapons systems became a preferred method of recycling petrodollars, with the bonus of acquiring instant prestige by owning systems found only in the inventories of the major powers.

*The perception emerges that arms exports negatively affect the economies of Third World recipients.* For the first time, there is an established linkage between recipient spending for arms imports and a decline in economic performance.[52] In line with the larger North-South economic debate, studies began to show that the arms trade was responsible for negative economic consequences in some developing countries. Unveiled during this period was the Soviet practice of bartering arms for commodities such as bauxite, cotton, and fish. They would then sell (or "dump") all of these commodities on the market for immediate cash, thereby lowering the future price and causing longer term economic damage to the arms recipient. Within the United States, the Congress began to pass laws prohibiting arms sales to developing countries who had no need or national security justification for them. In effect, the international economic system began to influence arms trade patterns.

*Licensed production and other forms of coproduction do not lead to increased ability of the recipients in the system to produce independently and cost-effectively those weapons systems needed for critical national security needs and increased independence from the major suppliers.* The increase in the use of licensing, coproduction, and offsets between advanced producers and the developing world did not result in the promised movement from consumer to producer for most of these states. There was a great deal of attention to this linkage on the part of the governments of major suppliers, who feared that this mode of production would erode their privileged position in the system. They acted accordingly by restricting the transfer of those parts of the production process (such as technology) that might make their fears a reality. At the recipient level, despite much euphoria and ballyhoo, very few states made much progress. The presence of petrodollars in the system meant that many states simply paid cash for end items. Other states began to discover that their lack of infrastructure made the route to national industrialization through arms production difficult indeed.

**Arms Trade Control.**

*Perceived and actual negative consequences occur as a result of the expansion of the arms trade system.* The control of the arms trade remained at the national level during this period. Very little private and commercial trade existed, except that which was either tightly controlled by governments or at low levels of military capability. However, the pressures to export cited above led to a decrease in the concern on the part of the suppliers for the negative consequences accompanying the increase in trade.

The relatively sudden expansion of arms exports from the major suppliers created significant concern in the system, especially regarding exports from the United States. As mentioned previously, the rebirth of the study of the arms trade coincided with this international concern as the Western states, led by the United States, tended to sell first and ask questions about negative consequences later. This was all reinforced by the decline in national security as a demand factor in the recipient states when compared to prestige and power. And in the United States there was no shortage of concern about the increase in Soviet arms exports, although the USSR continued to supply arms in the more traditional way, with credits and gifts to recipients who were needed for the larger East-West struggle. Combined, there was a concern in the system that increased arms exports were out of control, and that the unilateral approach to restraint was beginning to fail.

Although it is a contentious issue, there is some evidence that these fears were not unfounded. By 1976, congressional hearings in the United States had revealed enough evidence in this respect that Congress passed a series of new laws creating procedures designed to reassert control at the national level. There was significant analysis that led to the conclusion that the export of advanced weapons under the current norms of the system would produce instability. The outbreak and conduct of the 1973 war in the Middle East is a good example. Of the conflicts that did occur after 1973 (Angola, Lebanon in 1975, Iran-Iraq, and so forth), evidence emerged that they would not have occurred when and how they did without this expansion of the arms trade.[53]

*The control of arms trade remains at the national level as the continuance of the cold war prevents any formal multilateral controls.* The experience of the Carter administration in attempting to address these perceived and actual negative consequences tells a great deal about the arms control norms that existed during this period. One of the pillars of this policy was to engage the USSR in restraining the arms trade. Although several meetings were held, the talks were discontinued in the face of insurmountable differences when it came to what systems would be restricted and to which recipients.[54] Despite this failure, both the United States and the USSR continued to restrain their exports and provide what passed for arms trade control in the system, especially when it came to controlling conflict (see previous regime rules on conflict).

*Non-Soviet arms suppliers resist attempts to develop multilateral arms trade control regimes as they strive to survive as major arms producers through exports.* The new economic realities of the arms trade system created even more pressure against multilateral controls. As part of the Carter policy, the United States attempted to draw other suppliers into the talks, particularly the Western Europeans. The response was frosty at best, and in the case of France, the subject was not even allowed to appear on the agenda of international meetings. States determined to maintain their sovereignty by producing—and, by necessity, exporting—advanced military equipment had no interest in controlling the arms trade. The failure of the U.S.-USSR talks in December 1979 only added to the skepticism of the Western European states.

*Recipients resist any notion of multilateral arms trade control, viewing it as a major threat to the primary goal of national security.* The United States also attempted to draw the recipient states into the effort to control the arms trade. At this level there was even more resistance, as this was seen as an attempt by the "northern" states to infringe on the one area of policy in which they remained somewhat independent, the ability to defend themselves. Attempts in the United Nations met with a similar response.[55]

*Arms embargoes as a means of multilateral arms control are not successful.* Although the two superpowers continued to practice some restraint in regard to specific weapons systems and clients during this period, the several multilateral arms embargoes that were implemented during this period were failures. In the case of South Africa, an uneven implementation enabled it to develop enough indigenous defense production to maintain its universally condemned system of apartheid, as well as defend its territory in a more conventional military sense.[56] Similar results occurred when the United States restricted arms supplies to Brazil and Chile in this period.

Table 5-18 summarizes the 1966-80 arms trade system.

**Table 5-18**
**Summary of 1966-80 Arms Trade System**

| | |
|---|---|
| *Boundaries and Environment* | • 1966-1980<br>• Loosening of bipolar political system leads to intrabloc arms trade rivalries<br>• Continuing East-West conflict shifts from Europe to developing world and continues to provide major basis for arms trade<br>• Economic recovery of Western Europe means more suppliers<br>• Decolonization and petrodollars mean more recipients<br>• New international economic system, surfeit of petrodollars, and increased cost of advanced military technology creates economic incentives to export |
| *Characteristics of Units* | • Suppliers are national governments<br>• Multinational corporations emerge |

*Table 5-18 continued*

| | |
|---|---|
| | • Ratio of exports to production increases dramatically<br>• Recipients are primarily developing countries, increasingly unconnected from the East-West conflict |
| *Structure and Stratification* | • Arms transfers increase by 400 percent<br>• Number of suppliers doubles<br>• Thirty-three new recipients added<br>• U.S. and USSR still major two suppliers, but European and Third World producers increase their share of the market<br>• Increase in multiple supplier-recipient relationships<br>• Regional shift from Europe and Asia to the Middle East<br>• Proliferation of advanced weapons systems |
| *Modes of Interaction* | • Cash and credit become the dominant modes of payment<br>• Licensed production to Third World recipients<br>• Coproduction within East and West blocs<br>• Multinational production |
| *Regime Norms and Rules* | |
| Conflict | • The conduct and outcome of interstate regional conflict is determined primarily by U.S. and Soviet arms transfer policy and behavior.<br>• The control of the outbreak of interstate conflict by arms suppliers, especially the superpowers, is very difficult given the proliferation of military capability in the system.<br>• Intrastate conflict and terrorism is not affected by the arms-trading system, as systemwide proliferation of low- to middle-level weapons systems allows both government and oppositionist participants to use multiple sources for weapons, including the private commercial market.<br>• Third World recipients of modern weapons systems do not have the quantity, infrastructure, or manpower skills required to use these systems effectively in armed conflict or to threaten the major powers militarily.<br>• The presence and sudden availability of first-line military equipment, attributable to systemic factors of bloc obsolescence and rising oil prices, creates significant demand in the developing countries for these weapons systems for prestige purposes. The threat of armed conflict is no longer a prerequisite for receiving or exporting modern military equipment. |
| Foreign Policy and Diplomacy | • Arms transfers used as a bargaining tool to gain political influence continues to dominate as a rationale for the major suppliers, particularly in regard to gaining strategic access in the Third World.<br>• Arms transfers are not a reliable bargaining tool for influencing the short-term behavior of clients in a system where recipients are acquiring arms less for military reasons and more for prestige and regional political power, especially when they can spread their dependence among several suppliers, develop indigenous systems, or accomplish national security objectives with less modern equipment.<br>• Arms transfers result in structural influence of the patron over the client's medium-term behavior and ensure stability in the dominant supplier-recipient relationships. |
| Economic Effects | • Major arms-producing states must either export a significant percentage of their production of advanced systems or expire as an independent actor. |

## Table 5–18 continued

| | |
|---|---|
| | • Economic pressures make major suppliers, and particularly their arms industries, more likely to export technology and other aspects of the production process in order to survive as independent producers of advanced equipment.<br>• The sudden acquisition of convertible cash pressures many recipient states and their allies to recycle petrodollars by acquiring very advanced and expensive weapons systems, exacerbated by the "prestige race" that ensues. Poorer states with greater security needs tend to be less preferred by suppliers needing the cash.<br>• The perception emerges that arms exports negatively affect the economies of Third World recipients.<br>• Licensed production and other forms of coproduction do not lead to increased ability of the recipients in the system to produced independently and cost-effectively those weapons systems needed for critical national security needs and increased independence from the major suppliers. |
| Arms Trade Control | • Perceived and actual negative consequences occur as a result of the expansion of the arms trade system.<br>• The control of arms trade remains at the national level as the continuance of the cold war prevents any formal multilateral controls.<br>• Non-Soviet arms suppliers resist attempts to develop multilateral arms trade control regimes as they strive to survive as major arms producers through exports.<br>• Recipients resist any notion of multilateral arms trade control, viewing it as a major threat to the primary goal of national security.<br>• Arms embargoes as a means of multilateral arms control are not successful. |

# 6
# The Declining Bipolar Arms Trade System, 1980–92

## The Systemic Contradictions of 1980–85

Describing the international arms trade system that evolved after 1980 is more difficult than doing so for the previous historical systems. There are many contradictions when viewing the political, economic, and military dimensions of the larger global system. And it is difficult to set the temporal boundaries of the system. It is the familiar problem of continuity and change: in the final analysis, the boundaries must reflect one of the basic premises regarding the utility of systemic-level analysis set forth in chapter 3. In other words, the new system (change) must be defined in such a way that it explains the arms transfer behavior of its units more clearly than if the former system were assumed to remain in place (continuity).

If one reads assessments of the larger international system made in the middle of the 1980s, it will be noted that enough system-level shifts had occurred to be confident that a system transformation had taken place, particularly when one sees that the actors in the system were forced to respond by adjusting their arms trade behavior. As will be seen, this was particularly true with the significant changes in the international economic system, which forced arms suppliers to adapt with creative financing. One would also observe that the overall levels of arms trade to the Third World began to decline in the early 1980s after a steady climb upward during the 1966–80 period. On the other hand, the momentous changes in the international political system did not begin until 1986 and took shape only late in the decade. As seen in the conflicts in Nicaragua, Angola, Ethiopia, and Afghanistan, the two major arms suppliers (the United States and the USSR) could and did delay their response to this systemic change. But by 1992 the cold war had ended and the Soviet Union has disintegrated into autonomous republics, both events having a major impact on international arms trade. As mentioned earlier in this book, system boundaries are somewhat artifactual and can only be judged by their utility. All things considered, it will be demonstrated that the period from 1980 to 1992 should be considered as a distinct arms trade system, since the actors in the system were forced to adapt to pressures quite different than those in the previous bipolar system.

There were many signs that system transformation was under way by the early 1980s. The explosion of petrodollars had been spent (in many cases, misspent) and replaced by an international debt crisis as the price of oil dropped radically. Ironically, the resurgence of U.S. defense spending, financed by foreign borrowing, exacerbated this decline in the economic growth of developing countries.[1] The total magnitude of arms transfers to developing countries began to fall in the early part of the decade. The United States and the Soviet Union, who had armed Iran and Iraq, were no longer the dominant suppliers now that their clients were at war. Brazil, a net importer of armaments just a decade earlier, became a major supplier in the system. Genuinely commercial arms trade, including some that was illegal, resurfaced after an absence of more than forty years. Practices such as using offsets put a much more commercial slant on the arms trade.

By the end of the 1980s, sweeping changes in the larger international political system put the interim period of 1980–85 in proper perspective. The arms trade system had developed a more cohesive structure and set of regime rules that were linked to the early 1980s. It influenced and governed the behavior of the actors in the system through 1991 and up to early 1992 when the system had clearly been transformed.[2]

## Boundaries and Environment

The evolution of the international political system during this period had a significant impact on the arms trade system. For half of the decade, the cold war actually heated up. The Reagan administration was elected based on a platform of seeking to regain ground lost to the USSR during the post-Vietnam era. This renewal of bipolar competition had actually started in the last year of the Carter administration as a result of the Soviet invasion of Afghanistan. As a result, on many issues the conflict between the USSR and the United States returned to a level not seen since the early 1960s. Increased emphasis was placed on righting the perceived imbalance in nuclear weapons, which would include introducing the Strategic Defense Initiative in the United States. There was also a renewed effort to challenge the USSR and its clients in the Third World, in places like Lebanon, Grenada, Afghanistan, Angola, and Nicaragua. The Soviet Union responded in kind, and the system looked very much as it had in the past.

But with the ascent to power of a Soviet leader in 1985 who clearly wanted to tone down, if not eliminate, the East-West military tension, the political system began to shift significantly. The Reagan effort to boost arms exports in the name of a "strategic consensus" against the USSR in the Third World never really caught on, and by 1985, U.S. exports were in decline. As the rest of the world saw the East-West conflict on the wane, it began to act in a much more independent fashion. This included the West European members of NATO. Once the threat to Europe began to decline, almost certainly by the time the USSR and the United States agreed in 1987 to destroy their medium-range ballistic missiles in Europe, the NATO countries were much less interested in the so-called Nunn Amendment international armament cooperation programs with the United States.[3] This declining threat coincided with the decision by the European Community in 1985 to move toward a single market by 1992. By the end of the decade, the Conventional Armed Forces in Europe (CFE) arms control agreements made it clear that the 1990s would see significantly lower defense budgets and arms production in the industrialized world.

Combined, these major political shifts created movement toward an integrated European market in defense systems, with significant implications for the arms trade.[4] By 1990, the Soviet Union had set its former allies in Eastern Europe free and was in the process of dismantling its once-large overseas empire. At the July 1991 G7 meeting, it became clear that the superpowers were no longer dominating the agenda of the international political system. The fact that the U.S.-USSR summit was held *after* the G7 meeting only confirmed this reality.[5]

When the 1986-89 period is compared with that of 1982-85, arms transfers had declined in seven of the third world's ten leading recipients, with only Soviet clients—Vietnam (seventh), Afghanistan (eighth), and Angola (ninth)—receiving increased amounts.[6] As the former Soviet economic system continues to unravel, it is doubtful that even these recipients will continue to receive this level of support. The USSR was unable to sustain its grant aid and low interest credit programs to former recipients such as Nicaragua and Cuba. In short, by 1990, the decline of the bipolar system and its patron-client patterns was in full swing and was reflected in the arms trade system.

The end of the cold war has led to a significant and well-documented decline in the defense budgets of the major industrial states; most assessments forecast a net decline of 25% by 1995. This is having several important effects on the current international arms trade system. There has already been a significant loss of jobs in the military industries and a general decline in defense business, which has increased the pressure to "export or expire." In the Western democracies, this has also led to declining public support for defense industries and arms exports in particular. In the short run, the decline of the bipolar system has removed the primary rationale for defense spending, although the extent to which this is true varies from state to state.

The regional conflicts that dominated all of the postwar era continued during this period. There was no shortage of wars requiring the importing of conventional arms. As the decade came to a close, there was a decline in the number of conflicts requiring large-scale acquisitions, and the UN became active in conflict resolution. But it was premature to declare that this downward trend in global conflict represented a systemic shift. The Iraqi invasion of Kuwait, and other conflicts that continue unabated, indicate that the level of conflict will remain similar to previous systems. Additionally, during this period the dominant mode of conflict completed its evolution from the tightly controlled bipolar system to one in which regional conflicts are conducted based more on local grievances and capabilities, and are only partially affected by major-power interests and capabilities. The first major conflict of this period, the Iran-Iraq war, was the first of the postwar era in which the United States and the USSR failed to influence significantly the outbreak, conduct, and outcome of the war. Neither participant had indigenous defense industries, at least in the first few years of the war. With the USSR initially holding back on its exports to Iraq, and the U.S. embargo of Iran, the war was greatly influenced by arms imported from nations other than the superpowers. As with previous systems, there was no multilateral arms trade control regime, ensuring that in this system, regional conflict continued to attract arms.

Economically, this decade saw the end of the centralized economy as a competing system to capitalism and free markets, prompting some to refer to "the end of history."[7] Although the speed of the reform movement toward capitalism varies significantly from country to country, there is no mistaking the fact that the socialist approach had been seriously discredited. The arms trade system kept pace with this evolution, as will be seen with the shifts toward more commercial modes of transfer and production. With their potential to bring about rapid and significant negative consequences, arms remained a commodity requiring careful treatment by governments. But the arms trade system was significantly affected by the move toward private-sector economics.

The bulk of the arms traded continued to be exported to the developing world. As a result, the international debt crisis and decline in GNP growth that gripped the developing world throughout the decade had a major impact on the arms trade. Although there have been some studies that link this debt to the expensive arms acquisitions of the previous period,[8] in the main this economic development was systemic in nature, caused mainly by overspending in general and a major drop in oil prices. One effect of this change is that the reduction of the aggregate world military production base is lagging behind the political shifts noted above. Therefore, the decline in transfers to the Third World has been accompanied by a relative increase in those to the industrialized world.[9]

It has now become clear that the shortage of money on the part of the recipient actors has spread to the system at large. Once the euphoria of the changes in the international political system had died down, the world was left to contemplate a system seriously lacking in the capital required to correct the damage done by forty-five years of communist rule in eastern Europe and the aftermath of both the Iran-Iraq and Persian Gulf wars. "A new study by Morgan Stanley & Company said the total demand for capital by Eastern Europe, Latin America and the Middle East would exceed the Western world's supply by more than $200 billion a year in the next few years."[10] Nowhere in this study is there mention of any of this capital being needed for military projects. In such a tight economic environment, actors in the international economic system are openly arguing that available money not be spent for military purposes. One target of this attack is OPEC, the charge being made that the huge oil profits are responsible for the cycle of violence in the Middle East.[11] Both the World Bank and the International Monetary Fund have openly called for linking financial aid to a reduction in defense spending by recipient governments.[12] Germany recently announced that it will soon begin cutting its foreign aid to countries that spend too much on weapons.[13] The trends are too clear to avoid concluding that a troubled international economic system is causing a major downturn in spending for arms acquisitions.[14]

The recent report by the congressional Office of Technology Assessment, *Global Arms Trade,* stated that the most important macroeconomic force acting on Western defense industries was the decline in military expenditures and procurement that began in 1987. The most important microeconomic force was the "rapidly rising costs associated with weapons research, development and production."[15] Both of these systemic economic realities have begun to result in a serious decline in the arms trade, especially in modern major weapons systems.

This was not a period of rapid change in military technology, particularly when compared with the previous system, when a definite shift occurred from the immediate postwar period. The advances that did occur tended to be related to the U.S.-USSR military balance, were very expensive, and were incorporated into weapons produced only by the most industrialized countries. Producing these systems remained a very expensive proposition, and many would-be producers of the previous system (the second-tier suppliers) found that they were unable to keep up. Likewise, recipients had a more sober view not only of the expense of these systems but also of the infrastructure and skills required to utilize this advanced technology effectively. As will be shown, there was a definite trend toward the modernization and upgrading of current systems with advanced components, and away from buying the most modern and high-technology airframe, tank, or ship.

## Characteristics of Units

The major actors in this system continued to be national governments, as the transformation that occurred more than 50 years ago toward governmental control of military production and exports continued to hold. Armaments, particularly major systems such as tanks, aircraft, warships, and missiles, continued to be viewed as special commodities with significant political and military implications. And the individual rationales of the major suppliers—political, military, and economic—remained basically the same in this system. The "export or expire" norm remained a characteristic of all suppliers in the system, as seen in an even higher ratio of exports to overall defense production in the face of declining defense budgets. This occurred simultaneously with a decline in arms trade as a percentage of overall trade (see figure 6-1).

One effect of this reality is that despite the rise in importance of new types of nongovernmental actors in the system (see below), problems associated with the negative consequences of the arms trade are still addressed by governments and not by private defense firms. As defense budgets declined in this system, it was governments who continued to attempt to control and promote arms transfers to achieve national economic, political, and military objectives.

This is not to say that there were not significant actors who entered the system. Military industries and trading firms, which heretofore behaved almost exclusively as adjuncts to or subsidiaries of national governments, have gained enough independence to warrant the increased attention of analysts monitoring the arms trade. This is a direct result of the evolving international economic system described above, the move away from state-controlled enterprises. Not since the interwar period has so much interest been paid to describing and understanding how defense firms work in this new environment.[16] One example is the issue of ballistic missile proliferation. At one level, it is a familiar problem to analyze: the Chinese delivery of ballistic missiles to the Persian Gulf and the Middle East is opposed by the United States. The U.S. secretary of state engages the Chinese government in talks, in the hope that traditional diplomacy, perhaps coercive in nature, will stem the flow.[17] At another level, however, the traditional approach is no longer applicable. When West German firms were suspected of assisting Iraq and several other countries in developing the Condor II ballistic

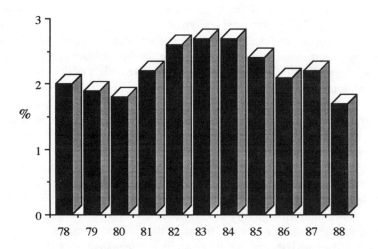

Figure 6-1. Arms Exports (Deliveries) as a Percentage of Total Exports

Source: U.S. Arms Control and Disarmament Agency, *World Military Expenditures and Arms Transfers 1989.*

missile, intergovernmental approaches proved inadequate. The reason was that the negative consequences resulted from the behavior not of national actors but of private firms, in most cases acting within legal limits set by the West German government.[18] This example also points out the basic weakness of setting up intergovernmental regimes such as the Missile Technology Control Regime (of which West Germany is a member) in a system in which it is getting more difficult to monitor the activities of private firms.

With the rise in importance of defense firms as actors, the system also saw the presence of truly illegal and private arms trade for the first time since the interwar years. Arms exports are *illegal* in cases when, if the facts were known at the time of the transaction, they would violate national arms export laws or the laws of the recipient country. A *private* export is one that is nongovernmental and hidden from the rather extensive bureaucratic process designed to control arms exports. In addition, this type of transfer involves technology and weapons suited mainly for conventional warfare, as opposed to the strategic systems that have been the focus of the effort to control the flow of technology from West to East. Though much of this illegal activity was spurred by Iran's attempt to keep its U.S. arsenal running without official U.S. support,[19] the phenomenon was global in nature.

There were enough celebrated scandals in this period to convince most analysts that new actors were now in the system.[20] At one point, two editorials appeared in *International Defense Review,* a respected journal that serves as an indispensable outlet for advertising and news of recent military developments for industries and governments involved in the international arms trade. Entitled "De Facto De-regulation of the Arms Trade" and "Some Rules of the Road for the Arms Traffic," they expressed concern that "unscrupulous arms merchants," "shady intermediaries," false end-user certificates, and the increasing use of those by governments was resulting in the "indiscriminate spread of arms" throughout the world.[21] Although the magnitude and effect of this trade remains a contentious issue,[22] there is little doubt that it had become a pervasive element in the system. Stories of the arming of Iraq uncovered illegal networks and transfers. The BCCI bank scandal in 1991 in which a very large international bank was discovered to have been used extensively for illegal arms sales, confirmed that illegal arms trade was big business.[23]

In the 1970s, Third World or second-tier arms-producing states entered the system as international arms suppliers. The major countries in this category included Brazil, Israel, a revitalized China, Spain, North and South Korea, Egypt, and South Africa. By the early 1980s, they were having a significant impact on the system, at least in terms of the amount of attention they were receiving from analysts and the other actors in the system.[24] By the end of the decade, two things had become clear about these new actors. First, they had found it very difficult to compete with the major producers (the superpowers and Western Europe). The debt crisis had slowed economic growth to the point where many of the industries in these countries, which just five years earlier were booming, were experiencing great difficulty and in some cases disappeared. They also found that success at transferring mid-level technologies to countries with an immediate military need was not a stepping-stone to the production and export of weapons systems with more advanced technologies. Almost all of these actors became dependent on key technologies provided by the major industrial states. Although some advanced systems were coproduced in these second-tier countries, their export was controlled by major-power governments. The 1988 CRS study provides further evidence of this rise and fall. The data in table 6-1 reflect arms transfer agreements with the developing world, measured in millions of current dollars.

But a second effect of the second-tier suppliers had become clear, and that was their ascendance as suppliers to those states engaged in interstate conflict. This could be foreseen at the time of the second-tier countries' arrival on the scene in the early 1980s. As a group, they tended to supply equipment that was battle tested and worked well in combat. It was generally lower in price[25] and could be built and/or packaged to suit the specific needs of customers. This also meant that, unlike arms from the major suppliers, the equipment and its accompanying tactics and doctrine did not come designed for the European theater. Salesmen from these second-tier suppliers were able to exploit the fact that both supplier and recipient were fellow members of the Third World, thus providing moral ammunition for the larger North-South conflict.

One effect of emphasizing this North-South rationale was an increase in second-tier suppliers engaging in cooperative programs without the support of the major powers. None of these second-tier suppliers were around in the 1930s, when the major-power suppliers agreed to control arms at the state level, and therefore they did not have the same aversion to being labeled "merchants of death." As a result, their products came with fewer strings attached. Most noticeably absent was the dreaded end-user certificate used by major suppliers to control not only how and against whom the weapons would be used, but also their reexport for economic and political gains. The 1990 SIPRI analysis of the arms trade (see table 6-2) concludes that "the significance of these countries as suppliers of countries at war is far greater than their overall importance within

**Table 6-1**
**Ranking the Second-Tier Suppliers, 1980-87**
*(millions of current dollars)*

|  | 1980-83 | | 1984-87 | | Percentage Change |
|---|---|---|---|---|---|
|  | Values | Rank | Values | Rank |  |
| Spain | 2,030 | 9 | 1,805 | 9 | −11.08% |
| Brazil | 1,890 | 10 | 1,655 | 10 | −12.43% |
| South Korea | 2,360 | 8 | 910 | 11 | −61.44% |

Source: U.S. Congressional Research Service, *Trends in Conventional Arms Transfers to the Third World by Major Supplier, 1980-87*. (Washington: Library of Congress, August 1988).

### Table 6-2
### Exports by Selected Suppliers to Countries at War as a Percentage of Total Exports of Major Conventional Weapons. 1980-89

| | |
|---|---|
| Syria | 99 |
| Libya | 96 |
| Egypt | 90 |
| Brazil | 47 |
| China | 40 |
| USSR | 35 |
| France | 23 |
| United Kingdom | 9 |
| United States | 5 |
| West Germany | 2 |

Source: SIPRI Yearbook 1990.

the global market place."[26] The other side of this correlation is that with the end of the Iran-Iraq war, these suppliers have had to await a new war of similar magnitude. Some, such as Brazil, are finding it very difficult.[27]

A third impact of the presence of these second-tier suppliers in the system was that an increasing amount of their production was consumed indigenously, thus lowering the demand for and levels of imported equipment. This is especially true for those mid- to low-level technology items that the major suppliers no longer produced. SIPRI's recent study of indigenous electronics industries in the Middle East suggests that these systems are becoming more important elements of regional military capabilities.[28] In sum, one conclusion that could be reached is that at the systemic level, although *demand* for military capability has increased as measured by larger inventories of equipment and personnel, the *trade* in military equipment has declined because of a parallel increase in the number of countries now able to produce the equipment indigenously.

Multinational corporations, albeit still controlled by national governments, continued to play a role in the system. The success of Panavia in exporting Tornado fighter and air defense aircraft to countries outside of the three producers of the aircraft (for example, Saudi Arabia) moved multinational arms exports to a higher plane, with some important implications for the system. First, the dollar value involved was significant, and this trade literally kept several major aircraft manufacturers in business. Second, it provided the example for other European multinational corporations and projects that got under way during the 1980s. Eurofighter, the consortium that is developing the European Fighter Aircraft (EFA), was modeled after Panavia. Also, the current efforts among Western European defense industries to develop "Europe only" cooperative programs—and perhaps exports—rely on the experience of the 1980s for lessons learned and encouragement. Third, this trade provided a mechanism for arms suppliers with restrictive declaratory export policies to reap economic benefits while avoiding domestic political opposition to expanded sales. West Germany's export policy did not allow export to "areas of tension," and as result, it had missed out on the economic benefits from selling to oil-rich countries such as Saudi Arabia. However, this changed in the case of the Tornado, when in 1983 West Germany went along with the other two members of the consortium (the United Kingdom and Italy) in approving a change to the procedure whereby only a majority vote was needed to make export decisions. As a result, when the United States balked at sending more F-15s to Saudi Arabia in 1986, the Tornado was sold instead. Although listed in most statistical tallies as a British sale, the West German share of the profits was significant.[29] A final implication of the rise in multinational arms trade relates to its impact on information of the system as whole. To the extent that this type of activity becomes more prevalent, traditional tracking and analysis of the arms trade, and particularly its consequences, becomes more problematic.

As for recipients as actors in the system, some of their characteristics remained similar to those of the previous system. Most of the countries engaged in actual armed conflict still did not possess the indigenous production capability necessary to unhook themselves from the arms imports needed to conduct the war. As pointed out above, the pattern of who supplied these counties at war changed, but the reality of dependence on outside suppliers remained. A change that did occur was the disappearance of the wealthy recipient making massive acquisitions for purposes of prestige. Instead, the typical recipient, both developed and developing, internally adopted a model of military-industrial development whereby through offsets, arms acquisition deals were designed to develop the indigenous employment, technology, and infrastructure necessary to become less dependent on the major industrialized arms suppliers. As will be seen when the modes of interaction are discussed later, very few major arms deals now take place without offsets. Arms suppliers in this system faced a recipient equipped to deal not only with its own defense needs but also with the pressing requirements of economic and industrial development that will lead to increased independence.

## Structure and Stratification

The political, economic, and conflict environment of this period resulted in an eventual decline in the magnitude of the arms trade in the 1980s, no matter which data are used to illustrate the trend. ACDA data on global deliveries for all types of recipients and weapons (including ammunition and spare parts) show a clear decline starting in 1985. This decline also appears when agreements and deliveries to the Third World are plotted (see CRS data in figure 6–2). When the data are restricted to the delivery of major weapons systems to developing countries (see SIPRI data in figure 6–2), deliveries remained high, with the decline occurring in 1988 and 1989. In essence, although international systemic factors are causing a decline in global arms transfers, the Iran-Iraq war deliveries bucked the trend until the cease-fire in 1989. It should be noted that arms

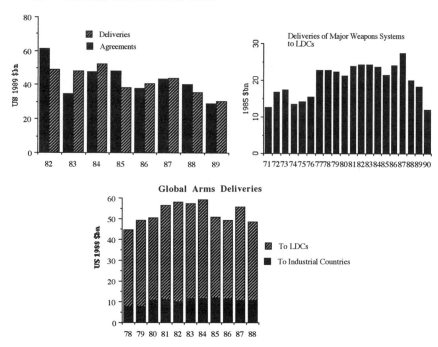

Figure 6-2. Arms Deliveries and Agreements in the 1980s

Sources: Above left, Congressional Research Service; above right, *SIPRI Yearbook 1991*; below, ACDA, *World Military Expenditures and Arms Transfers 1989*.

transfers declined to levels seen during the 1970s, still a very significant amount.

Recent ACDA data indicate that the ratio of developing-world to industrialized recipients remained basically the same, with an increase toward the end of the decade. In 1978, the industrialized world received 17.48 percent of the world's arms imports. In 1985, this figure peaked at 23.5 percent, falling to 22.2 percent in 1988. SIPRI's data, which tracks only major weapons systems, finds an even more pronounced shift toward industrialized recipients.

> The share of the industrialized countries in the global trade was approximately 33 per cent in 1987, 42 per cent in 1988, and 50 per cent in 1989.... The gradual but constant growth of imports of major conventional weapons systems by industrialized countries during the period 1985-89 reflects the rising importance of Japan and—despite the prospects for conventional arms control in Europe—the NATO countries.[30]

This suggests that the shift will be short-lived and represents more of a last gasp prior to the system resuming its 80 percent/20 percent balance in favor of the developing world. The SIPRI data published in June 1991 indicate that the "trend of a shrinking Third World share in the world arms market of major conventional weapons that has been observed for several years was not continued in 1990," with this share remaining at the 1989 figure of 45 percent.[31] Poland and Czechoslovakia, major industrialized arms importers in the 1986–90 period, were no longer clients of the USSR.

*Suppliers*

There was little change in this period in the number of suppliers in the system (table 6–3), especially in the number of suppliers that mattered. The system appeared to have peaked on this dimension. There was no shortage of observers predicting that if the number of suppliers remained constant and the demand dropped, some suppliers would drop out. Although it will be seen that transfers from some suppliers declined drastically, the role of governments in arms production and trade has prevented the diminution in the number of suppliers one might expect in a purely economic market system.

*Recipients*

The number of recipients in the system also leveled off, as indicated in table 6–4.

### Table 6–3
### Number of Suppliers in the System, 1975–88

|  | 1975 | 1980 | 1985 | 1988 |
|---|---|---|---|---|
| Suppliers with exports to developing countries > 0.1% of total (SIPRI data) | 20 | 20 | 20 | N/A |
| Suppliers with exports to all countries > $10 million (ACDA data) | 33 | 34 | 39 | 43 |

### Table 6–4
### Number of Recipients in the System, 1975–90

|  | 1975 | 1980 | 1985 | 1990 |
|---|---|---|---|---|
| Number of recipients on SIPRI arms trade register |  |  |  |  |
| Industrialized recipients | 30 | 32 | 32 | 31 |
| Developing recipients | 80 | 86 | 85 | 74 |
| Number of recipients in ACDA's annual *WMEAT* publication (> $10 million) |  |  |  |  |
| Industrialized recipients | 29 | 31 | 31 | 31[a] |
| Developing recipients | 64 | 81 | 79 | 82[a] |

[a] 1988 data.

136 • *The International Arms Trade*

*Market Shares.*

There was some definite continuity in terms of market shares in this new system, whether one looks at just the developing world or the world as a whole. Figure 6–3 indicates that the United States, the USSR, and Western Europe still accounted for roughly 80 percent of agreements and deliveries with developing countries. However, by 1988–89 it was clear that Western Europe arms exports were in decline. When just major weapons systems deliveries to developing countries are considered, the overall percentage of the market held by these three is somewhat higher, and as of 1989, the relative USSR share was increasing. Given the environmental changes already described, this USSR share was bound to decline rapidly, and in fact it did so in 1990 and 1991.

The Congressional Research Service data released in August 1991 revealed that the United States had surpassed the USSR as a supplier of weapons to the

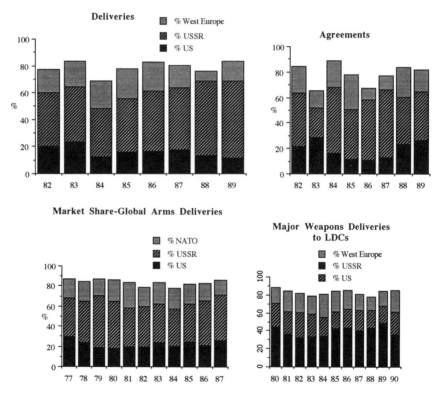

**Figure 6–3. Arms Deliveries and Agreements to the Developing World in the 1980s**

Sources: Above, Congressional Research Service; below left, ACDA, *World Military Expenditures and Arms Transfers 1989;* below right, *SIPRI Yearbook 1991.*

developing world. While U.S. arms transfer agreements totaled $18.5 billion for 1990, the USSR total was $12.7 billion. It should be noted that these numbers represent agreements, and it remains to be seen whether or not they take the form of actual deliveries. For example, the U.S. total for 1990 includes a $6.1 billion package of major equipment for Saudi Arabia, the delivery of which will be significantly stretched out and perhaps revamped downward, given the systemic factors outlined above.[32] Also, the normal relationship between agreements and deliveries may not hold in the 1990s, since it is not clear that Iraq will be paying off its debt incurred for agreements signed after the cease-fire with Iran. This may mean an exaggerated decline in deliveries in the next few years.

As previously discussed, the exports of second-tier suppliers peaked and then leveled off during this period. Depicted in figure 6-4 is a comparison between the ACDA data for these suppliers (deliveries to both developed and developing countries) and the SIPRI data, which include only deliveries of major weapons systems to developing countries. Both indicate a downward trend at the end of the decade, with very different peaks. The peaks in 1982 and 1984 may reflect the overall popularity of second-tier suppliers in a system that had become a buyer's market, to include industrialized recipients. At this time there was a growing use of offsets in the system, whereby second-tier countries bought advanced equipment from the industrialized world but the sale was "offset" by sales of components and mid- or low-level technology items to the industrialized supplier country. Transfers of trainer aircraft from Brazil to the United Kingdom and electronics from Israel to the United States are good examples of this phenomenon at work in the system.

The sudden surges in 1987 and 1988 (SIPRI data) coincide with the final and furious stages of the Iran-Iraq war, when agreements previously reached were finally realized through actual deliveries. That war illustrates how important

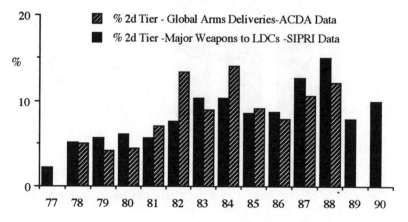

Figure 6-4. Market Share of Second-Tier Suppliers, 1977-90

these second-tier suppliers had become, despite the fact that their relative standing in the system, in an economic and technological sense, had peaked by mid-decade. The data in figure 6–5 show that despite an overall dominance in the system by the United States, the USSR, and Western Europe (roughly 80 percent) during the period, second-tier suppliers[33] played a dominant role in supplying the combatants to this war. The data also vividly demonstrate, in the number of countries who supplied both sides, that the system had indeed shifted to a buyers' market. This is especially true when compared to previous periods featuring a cold war political system that did not produce such a supplier-recipient pattern.

In sum, the second-tier suppliers became firmly established as *niche* suppliers, being available in those circumstances where the major suppliers were unable or unwilling to participate fully in the market. This may mean that a major supplier no longer produced those low- to mid-level technology items needed by Third World recipients. Or it may mean, as seen in the Iran-Iraq war, that international politics restricted participation by major suppliers. It should be remembered that the USSR instituted an embargo against Iraq for the first few years of the war, and starting in 1983, the United States instituted an embargo against Iran (Operation Staunch) that by 1986 was significantly restricting the flow of Western European arms to Iran.[34]

*Supplier-Recipient Dependencies*

Previous figures in this book have shown how dependency relationships (that is, the use of sole, predominant, or multiple suppliers) moved in the direction of

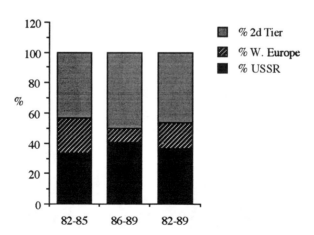

**Figure 6–5. Market Shares of Arms Deliveries to the Iran-Iraq War**

Source: Congressional Research Service (1990 data).

less dependence throughout the entire postwar period. The data in table 6–5 show that, at the beginning of the decade, there appeared to be a return to more sole- and predominant-supplier relationships in the system. That change, however, was mainly a reflection of Italy, Canada, South Korea, and the United Kingdom joining Japan and Israel as recipients that were 95 percent to 99 percent dependent on the United States for arms imports. But when the data are aggregated for the next two series of ACDA data (through 1984–88), the trend toward more sole and predominant relationships is unmistakable.

*Regional Distribution*

There were some changes in the regional distribution of major weapons systems to developing countries during the decade of the 1980s. Both the SIPRI and ACDA data in figure 6–6 depict a shift in 1988–89 to South Asia as the Iran-Iraq war wound down and arms sales to Afghanistan increased when the USSR departed (and both the United States and the USSR stepped up arms supplies in an attempt to see their clients survive). Also, India was the leading arms importer of the 1985–89 period, with major weapons acquisitions totaling $17.3 billion. However, recent SIPRI data for 1990 show a return to a system dominated by the Middle East. The data trends also show that arms transfers have declined in the three areas of the world hit the hardest by the debt crisis—Latin America, sub-Saharan Africa, and North Africa.

*Proliferation of Military Potential*

The previous system saw the exporting of advanced weapons systems to developing countries as a new phenomenon, albeit one that occurred mainly

**Table 6–5**
**Sole- and Predominant-Supplier Relationships, 1976–88**

|  | 1976–80 | | | 1981–85 | | |
| --- | --- | --- | --- | --- | --- | --- |
|  | 100% | 95–99% | 60–95% | 100% | 95–99% | 60–95% |
| United States | 1 | 2 | 16 | 4 | 6 | 16 |
| USSR | 7 | 5 | 21 | 3 | 4 | 20 |
| Other | 1 | 1 | 6 | 7 | 0 | 6 |
| Total | 9 | 8 | 43 | 14 | 10 | 42 |
|  | 1983–87 | | | 1984–88 | | |
|  | 100% | 95–99% | 60–95% | 100% | 95–99% | 60–95% |
| United States | 6 | 5 | 22 | 5 | 4 | 2 |
| USSR | 7 | 8 | 18 | 7 | 8 | 18 |
| Other | 7 | 0 | 9 | 2 | 1 | 8 |
| Total | 20 | 13 | 49 | 14 | 13 | 49 |

Source: U.S. Arms Control and Disarmament Agency; also Catrina, *Arms Transfers and Dependence*, pp. 6, pp. 388–91.

140 • *The International Arms Trade*

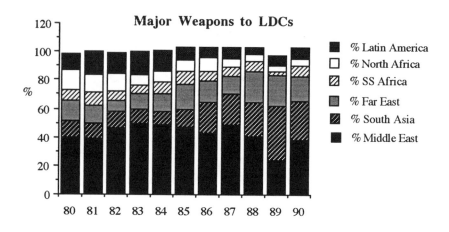

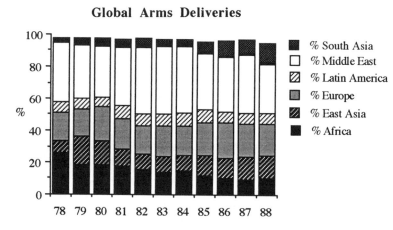

Figure 6-6. Regional Distribution of Major Weapons to LDCs and Global Arms Deliveries in the 1980s

Source: Above, *SIPRI Yearbook 1991*; below, *World Military Expenditures and Arms Transfers 1989*.

with wealthy recipients who wished to increase dramatically their political prestige. These recipients were few in number, and with the possible exception of Israel, they lacked the adequate infrastructure required to convert these advanced systems into enough usable military capability to effect a systemic change. In the 1980s, however, the export of advanced systems accelerated significantly and resulted in those states not directly involved in the East-West conflict (such as Iraq) acquiring what they perceived to be usable military power. At the end of the decade, when the East-West overlay began to crumble, these centers of military power stood as evidence of the new arms trade structure.[35]

Figure 6–7 from SIPRI's recent comprehensive study of the arms trade, depicts in the aggregate the proliferation of advanced systems.[36]

A review of the data reveals that an impressive array of countries now possess advanced weapons systems. But it is only when one takes a look at a particular type of military capability that the impact of this proliferation becomes clear. Although a thorough analysis of each type of capability is beyond the scope of this book, extensive studies have been done regarding the implications of at least three types of proliferation. The first of these is the proliferation of naval capability to the developing world, and its effects on the ability of the major powers to maintain their international interests through the use of naval power. In his book *Expansion of Third-World Navies*, Morris assesses the implications of this proliferation.[37] For example, regarding what he terms the "Third World naval hierarchy," there are both negative and positive arms control implications. The drive to be able to control exclusive economic

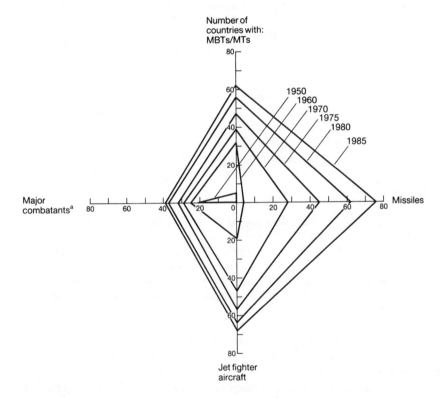

**Figure 6–7. Numbers of Countries in the Third World with Selected Weapon Systems, 1950–85**

Source: Brzoska and Ohlson, *Arms Transfers to the Third World: 1971–85*, p. 12.

zones may result in an "uncontrollable upward cycle of spending and even out-and-out arms race. . . . As aspirations about naval status escalate, there is a temptation to orient the national power base towards the support of naval expansion rather than just maintaining the military establishment for the purpose of protecting and insulating the national power base."[38] Positive implications include the fact that "strong navies, and therefore some degree of naval expansion, may be needed as a deterrent to potential aggressors in the often conflict-prone settings of the Third World."[39]

In Anthony's more recent definitive treatment of this aspect of the arms trade, he states:

> Events such as the loss of six British ships, including four major surface combatants, during the Falklands/Malvinas War of 1982 and the attack on the frigate USS *Stark* in the Persian Gulf on 17 May 1987 have highlighted the importance of air forces and air-launched guided missiles to maritime security. These incidents underline the rapid changes in the nature of naval weapon technologies that have taken place during the 1970s and 1980s. Not only have systems already in service increased their range, payload, accuracy and reliability, but a growing number of new weapon systems are now deployed by the navies of the world.[40]

Anthony presents data on the deliveries of submarines from 1946 to 1988, noting that sixteen states are in possession of modern diesel submarines. His review of recent air-launched antiship guided-missile deliveries shows that eighteen developing states acquired advanced ASMs such as the Exocet or Harpoon.[41] Anthony does caution that transforming these types of systems into usable military power, especially against the major world powers, is very related to the context.

> The understandable focus on events as dramatic as the Falklands/Malvinas War and the attack on the USS *Stark* may well have exaggerated the overall extent to which the gap between the naval capabilities of the major powers and those of naval forces in developing countries has closed. This is not to argue that this kind of missile proliferation is unimportant, but simply to point out that its importance does not stem from specific technical characteristics of the weapons themselves.[42]

Nowhere was this more true than in the most recent Persian Gulf War, where Iraq's navy was destroyed before it was able to perform even the most basic military action.

A very important facet of this type of proliferation is its linkage to changes in maritime law, specifically the expansion of offshore boundaries. As states opted for expanded economic and fishing zones out to 200 nautical miles, they began to acquire those surveillance and weapons systems that gave them some limited means to defend their territory.[43] In summary, the proliferation of weapons systems that result in increased sea denial capability in selected developing states is real and is having an impact on key naval powers. This is particularly true as the East-West naval rivalry declines and the major-power navies concentrate on preparing for those threats that remain. In March 1991, the U.S. head of naval intelligence testified before Congress regarding these new threats, indicating that the U.S. Navy had internalized a systemic change.[44]

A second area is the proliferation of advanced aircraft and surface-to-air missiles (SAMs), which combine to produce air defense capability. Ackerman's recent survey of the proliferation of air defense capabilities to the developing world shows that although this proliferation is concentrated in a few key areas, it is less context specific than naval sea denial capability and represents a much greater threat to the air forces of the major powers attempting to project power into certain regions of the developing world.[45] Tables 6–6 and 6–7 depict the type and magnitude of the proliferation of these systems.

Though a rational assessment might demonstrate that the proliferation of naval sea denial and air defense capability should concern the major powers and the world at large, the fact is that it generated little concern except in those cases in which it has been employed with negative consequences for the major powers. For example, in the recent Persian Gulf War, Iraqi SAMs and naval forces were eliminated in the first days of the war. Although this may mean that specific

## Table 6–6
### Aircraft and SAMs Supplied to the Developing World

| Aircraft | Country/Supplier | SAM | Country/Supplier |
|---|---|---|---|
| MIG-29 | USSR | Crotale | France |
| Tornado ADV | Panavia/U.K. | Shahine | France |
| F-16 | United States | Roland | Euromissile |
| F-15 | United States | Javeline | United Kingdom |
| F-18 | United States | Rapier | United Kingdom |
| Mirage 2000 | France | SA-13 | USSR |
| | | SA-14 | USSR |
| | | I-HAWK | United States |
| | | Stinger | United States |

Source: SIPRI arms trade registers, as compiled by Ackerman (see chapter note 45).

## Table 6–7
### Proliferation of SAMs and Air Defense Aircraft, 1980–89

| | 1980 | 1980–84 | 1980–89 |
|---|---|---|---|
| Number of AD aircraft | 220 | 976 | 1,661 |
| Number of recipient countries | 3 | 12 | 22 |
| Number of SAMs | 96 | 2,853 | 14,972 |
| Number of recipient countries | 1 | 8 | 18 |

Source: SIPRI arms trade registers, as compiled by Ackerman (see chapter note 45).

major powers may have to adjust to these new threats, the response would be a function of factors at the national level. And even in those cases, the issue is less the advanced weapons themselves than the larger political context. In sum, the proliferation of naval capability and air defense systems is seen by the actors in the system, for the most part, as a legitimate exercise of states' sovereign right of self-defense.

However, the proliferation of ballistic missiles in the developing world is a different matter. Developing countries possessing long-range ballistic missiles (ranges greater than 900 km) include India, Iraq, Israel, and Saudi Arabia. The celebrated Condor II will have a range of 1,000 km if it is ever developed; it was scheduled for acquisitions by Argentina, Egypt and Iraq. The following countries have ballistic missiles of a 200 to 900 km range: Afghanistan, Egypt, India, Iran, Iraq, Israel, Libya, North Korea, Pakistan, Yemen, South Korea, and Syria.[46]

This proliferation was met by significant attention and resistance by the major powers. There are several reasons for the development of this sizable international concern. First, there is no known defense by any power, other than preemption, against this capability, which is clearly offensive in nature. Efforts are under way to create a defense against this threat,[47] but it is not likely to become a reality in the short term. Second, most of the states acquiring these missiles are also seeking to acquire warheads of mass destruction. This inexorably links the issue to the nuclear nonproliferation regime. Third, this is a weapon system that many feel is inherently unstable, increasing the likelihood of conflicts that impinge on the national interests of the major powers. It is the one advanced conventional weapon system that prompted a systemic arms control response in the 1980-92 system—the Missile Technology Control Regime (MTCR)—with the United States, Canada, the United Kingdom, France, Italy, Germany, and Japan agreeing in 1987 to control and deny the export of this capability to any developing state. Although this regime lacks any enforcement mechanism and is not working as well as some might like, it is in effect and had formally added nine states to the regime by 1991 (Spain, Belgium, the Netherlands, Luxembourg, Denmark, Norway, Australia, New Zealand, and Austria). The USSR also agreed in 1990 to abide by the rules put down by the regime.[48] The Iraqi invasion of Kuwait and the subsequent embargo of Iraq by the MTCR signatories and others was a real test of how the international system would treat this threat, since Iraq possessed significant numbers of missiles that could have been launched, presumably with chemical warheads. It should be noted that the arms control approach is not the only evidence that this proliferation is taken as a serious threat to the system. The fact that the major powers coordinated and utilized their military power to remove this threat forcibly was even stronger evidence of its perceived importance.

The proliferation of ballistic missiles and its implications for international security is being extensively studied.[49] It has begun to attract the attention of an arms control community that has concentrated mainly on nuclear threats for most of the postwar period, and will be addressed later in this chapter and in chapter 7.

The above discussion has highlighted three major types of military power that have proliferated to the developing world and will continue to do so in the next decade. The question remains, however, as to how effective this power will be. What will be the effects on regional conflict, on the interests of the major powers, and on the system as a whole? At a minimum, in the continued absence of East-West conflict and in an international system with rules still evolving, states possessing these newly acquired capabilities will be experimenting with their application as they move to advance their national interests. As will be shown in the final chapter of this book, the major powers are beginning to behave as if these new capabilities *do* matter.

The studies of how the acquisition of advanced military equipment will change the international security system do note that much more than major equipment is required to increase military power. Although the major end items receive the lion's share of attention from those charged with monitoring arms transfers, the 1980s demonstrated that procuring tanks, ships, aircraft, and missiles is not the only route to acquiring usable military capability. The trends developed by SIPRI and ACDA confirm a decline in the arms market as defined mainly by trade in end items, both in dollars and number of systems. There is a great deal of evidence, however, that the number of transactions in components, spare parts, and upgrades increased significantly.[50] Several developments may account for this. Many of the countries that bought major weapons systems in the 1970s now needed the spare parts to keep them operational, a natural part of the absorption process. The debt crisis and lack of capital for the purchase of the most modern weapons led to the creation of trade in refurbished equipment. And as recipients switched suppliers, often under strained conditions, direct support of equipment in the inventory became a problem. Military and defense journals that cater to defense industries have been featuring articles for some years on the growth of this market.[51] In addition, information companies catering to international defense industries now hold seminars and publish special reports on the market.

Because this type of trade has become increasingly commercialized, data bases tracking this trade are much harder to construct. Up until 1988, the RAND Corporation was systematically tracking this phenomenon as it applied to the United States and to European countries. It concluded that the demand for expensive platforms (aircraft, tanks, and ships) was down, whereas the demand for subsystems for these platforms and for stand-alone complementary systems (radar, communications, and so forth) was rising: "The one broad generalization that we can make is that the future of the large, integrated European aerospace companies is problematic, whereas the specialized electronics and subsystem producers appear to be thriving."[52] As another example, there has been significant growth in the number of countries producing subsystems that enhance naval capability, as measured by the entries appearing in *Jane's Weapons Systems* (table 6–8).

The trade in these types of subsystems is difficult to track, given the fact that they are less visible, both in a military/intelligence sense and in a political sense. For example, many observers were surprised at the ease with which Israel defeated the Syrian air force in 1982, because military balance analysis tends to focus on end items (numbers of aircraft, missiles, and so on). But it was the use of electronic subsystems that enabled Israel to prevail, subsystems that by their nature are invisible. And if these subsystems are less visible, the enhancement of military capability through the acquisition of dual-use technology in commodities such as computers complicates things even further. Although this was mainly a problem for the major powers, it became clear that in the 1980–92 international arms trade system, more actors were acquiring military capability through this method. The efforts of the Western states to prevent the Soviet bloc and China from obtaining such technology (through attempts such as COCOM) are well-known and not normally associated with conventional arms trade. However, with the upheaval in East-West political relations, many of the barriers to trade in these commodities are falling as all states scramble to use dual-use technology as a force multiplier. This has resulted in a more open market, with significant if not yet specified consequences for the post-1992 system. We are seeing the passing of the time when industry could be neatly divided into military and nonmilitary compartments, and the result has been an increasing amount of strictly commercial trade with significant military applications. Another route to increasing military capability in a financially tight market is the acquisition of used and surplus equipment. As the 1980s evolved, this type of trade increasingly became a part of the structure.[53] Once before, early in the postwar period, the international arms market found itself with massive quantities of surplus equipment. But the superpowers were in tight control of that system. In the 1980–92 system, a similar surplus developed for several

**Table 6–8**
**Proliferation in Production of Naval Systems, 1979–88**

| Naval System | Number of Countries | | Number of Items | |
| --- | --- | --- | --- | --- |
|  | 1979 | 1988 | 1979 | 1988 |
| Naval SAM | 7 | 8 | 28 | 28 |
| Naval radar | 10 | 13 | 229 | 341 |
| Sonar equipment | 8 | 12 | 128 | 307 |
| Torpedoes | 7 | 7 | 61 | 53 |
| Coastal defense | 5 | 8 | 7 | 10 |
| Naval SSM | 9 | 12 | 17 | 30 |
| Fire control | 13 | 12 | 123 | 128 |
| Guns rockets | 14 | 12 | 88 | 98 |

Source: *Jane's Weapons Systems*. 1979–1988.

reasons. As with the postwar period, life cycles for weapons systems are finite. As wealthy recipients upgrade their military capability with newer versions of the end items procured during the oil price boom of the 1970s, the older versions became available. Secondly, the recently signed agreements in Europe (such as CFE) made available massive quantities of tanks, artillery, aircraft, and perhaps missiles.[54] We saw the USSR transferring a lot of its equipment from its former Warsaw Pact allies to beyond the Urals so that it would not be counted as part of its European arsenal and therefore will not be subject to destruction under the agreements. In the previous system, dominated by cold war politics, the dominant interpretation in the West would have been one that stressed duplicity and the buildup of a reserve for use in offensive actions. In the 1980–92 system, a competing interpretation was that the USSR wanted to preserve these systems for sale on the international market. Added to this stockpile of used commodities were those exported by suppliers who do not require an end-use certificate. The retransfer of Brazilian armored personnel carriers from Libya to Iran is but one example.

## Modes of Interaction

### Modes of Payment

The arms market has become much more commercial in nature, as measured by the steady increase in the percentage of the trade that does not directly involve the government as a contracting party. Part of the explanation is the steady move away from corporatism and from the state control of industries in general. In the United Kingdom, for example, the Thatcher government had a national policy of selling off government firms to the private sector; this has included some major defense firms. Compounding this effect was the rise in importance of spare parts, components, and the upgrade market, as mentioned previously. Though many large defense firms remained under state control, these firms steadily decreased in importance as the strictly commercial suppliers of components and subsystems became the important players in the trade. The data from the United States (figure 6–8) make the point that approximately 40 percent of U.S. arms deliveries were commercial in nature.[55]

U.S. data also reveal (figure 6–9) that whereas credit sales were an apparent remedy for the lack of cash in recipient countries during the first part of the decade, by fiscal year 1989, credit as a percentage of agreements was about equal with commercial deliveries as a percentage of all deliveries.

### Modes of Production

**Internationalization of Production.** The international arms trade system is now marked by the internationalization of production. For aircraft, this means that

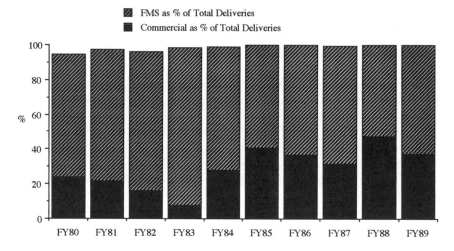

Figure 6-8. Trends in U.S. Arms Transfers (Commercial versus FMS), 1980-89

"aerospace industry corporations form joint venture multi-national companies, buy shares in foreign aerospace companies, buy and sell licenses to produce each other's systems, and use components obtained from various countries."[56] Introduced in the previous period with systems like the Jaguar aircraft, this type of production now has become standard.[57] The trend can also be seen in missile

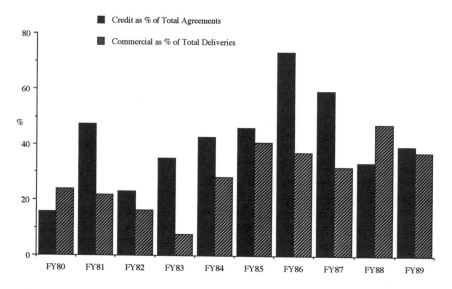

Figure 6-9. Trends in U.S. Commercial and Credit Arms Sales, 1980-89

production. The RAND database reveals that U.S. and other major NATO exports of platforms were down 20 to 30 percent in 1980–87. In contrast, NATO missile exports rose 20 to 40 percent, electronic sales tripled, and U.S. military electronic exports remained high. The data in table 6–9 on European missile projects make the point clearly.[58]

A brief look at the data on international programs contained in the various Jane's annual compendia of weapons systems provides a good summary of how the internationalization of military production became a reality of the arms trade system. In 1965–66, there were no entries in any book. By 1975–76 (as noted in the description of the 1966–80 system), aircraft and missile programs began to emerge; Jane's described six aircraft programs (Alpha Jet, Jaguar, MRCA, Orao, Concorde, and Airbus) and five missile programs (HOT, Milan, two versions of Otomat, and the Martel ASM) in about fifteen pages of text. By 1985–86, Jane's needed thirty-one pages just to describe the sixteen aircraft programs in place. International programs across all other types of weapons totaled eighteen.

Evidence on the movement toward multinational trade in naval equipment is provided in a major book by Sami Faltas, *Arms Markets and Armament Policy: The Changing Structure of Naval Industries in Western Europe*. Although it focuses on the naval industries of Europe, his conclusions conform to the larger international reality as well. His fourth chapter addresses the internationalization of warship equipment industries. He creates a data base that tracks the trade in the following categories of equipment for surface warships: antisubmarine-warfare (ASW) weapons, air defense weapons, surface-warfare weapons, underwater sensors, air and surface sensors, and main propulsion engines. In the early 1960s, industries in European NATO countries placed two-thirds of their products in national naval vessels, whereas in the 1980s they were exporting two-thirds of their output. Across the board, warship equipment industries in these states have become heavily dependent on exports, mainly to countries outside of NATO Europe.[59]

In the previously described international arms trade system, the Western European states, recently recovered economically and producing advanced military equipment for the first time since the 1930s, had their special place in the system. France, the United Kingdom, Italy, West Germany, and several other states were able to maintain a national production base through arms exports to

**Table 6–9**
**European Missile Trends, 1946–87**
*(number of missile projects)*

|  | 1946–59 | 1960–69 | 1970–79 | 1980–87 |
|---|---|---|---|---|
| National projects | 15 | 19 | 18 | 16 |
| Foreign participation | 6 | 11 | 21 | 31 |
| Total | 21 | 30 | 39 | 47 |

Source: RAND data base (see chapter note 52).

an oil-rich Third World and cooperative armaments programs with their NATO partner, the United States. Various efforts to integrate and rationalize European production (with the aim, for example, of producing one indigenous European tank, fighter aircraft, or air-to-air missile) always fell short. The United States, and many within Europe, were quick to point out that Europe was not up to such a task. It would be too expensive, and the final product would be technologically inferior to the Soviet equipment it was designed to counter. The economic and military benefits of weapons collaboration have been obvious since the 1960s. But as of the 1980s, not more than 20 to 25 percent of defense production in Europe could be characterized as collaborative in an international sense.[60]

The world of 1991 looked very different. The 1980s saw a steady decline in international demand for advanced weapons systems in the Third World. The events in Eastern Europe in 1989, along with a verifiable decline in Soviet activity in the Third World, saw a radical transformation of the bipolar system into an as-yet-undetermined international political structure. Whatever the empirical reality of this new system, the United States and the Western European nations have begun to respond to its central reality: they have significantly reduced their defense budgets in anticipation of the continuous reduction of the previous Soviet threat in Europe. Within Western Europe, changes are occurring that are not directly related to arms production and export. Since 1986, the European Community (EC) has been steadily moving, with a few fits and starts, toward a single integrated commercial market by 1992. Events in Eastern Europe and the Persian Gulf, and other non-European events, increasingly required a purely European response. The West European Union (WEU) was resurrected as a possible candidate for a strategic/military arm of the EC. The European members of NATO resurrected the long-dormant Independent European Program Group (IEPG) in an effort to integrate the military research and development activities of the European states.[61]

How did West European arms production change as a result of these larger systemic shifts? In previous, less dramatic periods of change, when faced with pressures to decrease defense production, West European governments tended to increase exports, pay more per item, increase efficiency through privatization, or in some cases drop production of a particular type of capability and acquire it from the United States or another supplier (for example, the United Kingdom's acquisition of the U.S. AWACS aircraft). In other cases, the Europeans have engaged in multinational production to solve the problem of insufficient national demand (for example, the Jaguar, Tornado, and European Fighter Aircraft [EFA]). The current systemic shifts, however, appear to be too large and permanent to allow an ad hoc response by the West European arms-producing firms and states. Commencing in 1987, the West European defense industries and governments began a serious effort to consolidate and rationalize defense production.[62] This effort has resulted not only in the consolidation of defense industries within national boundaries, but in a significant increase in pan-

European mergers, acquisitions, joint ventures, and other forms of cooperative defense production, all at the expense of transatlantic cooperation in armaments production. The data in table 6-10 may understate how this relationship has deteriorated.[63]

Within the EC, efforts are being made to include defense production and trade with the other commodities included under the Single European Act. Defense industrial cooperation between French and British firms proceeded at unprecedented levels. Though it is true that these developments are still very much in flux, that much national resistance remains, and that the full flowering of this trend will not occur until after 1995, there is no question that the movement toward European weapons systems for European military forces became well established.[64]

**Offsets.** One major change in the system that involves both a mode of payment and a mode of production is the rise of offsets, now a fact of life in making arms deals.[65] Recipient countries are now demanding that the price they pay for military equipment be offset by the supplier. This can be in the form of indigenous coproduction of parts and subsystems in the recipient country, technology transfer, or countertrade. Although this mode of interaction dates back to the early days of the postwar system, it was practiced mainly between the United States and its West European allies, either to aid in their economic and military recovery from World War II or as part of a growing effort to rationalize equipment inventories within the NATO alliance. For example, the F-16 deal in 1974, referred to at the time as the "sale of the century," was awarded to the United States based partially on its superior offset provisions. The European recipients—Norway, Belgium, the Netherlands, and Denmark—were unlikely to agree on either the French or Swedish aircraft also being offered, giving the United States a decided advantage. However, the recipients did want to maximize the potential for enhancing their own defense industries as a result of the purchase, and they succeeded in raising the amount of the offset as a result during the bidding wars.[66]

Table 6-10
Nunn Amendment Cooperative Research and
Development Funding, Fiscal Years 1986-92

| Fiscal Year | Amount |
|---|---|
| 1986 - | $100 million |
| 1987 - | $145 million |
| 1988- | $150 million |
| 1989- | $154 million |
| 1990- | $104 million |
| 1991- | $85 million |
| 1992- | $41 million |

Although offsets continue between major suppliers (such as the U.S. AWACS aircraft sales to the United Kingdom and France[67]), the practice has clearly gone beyond cooperation between the United States and industrialized recipients. In the current system, almost all recipients require that their purchase of arms be offset using a variety of arrangements, including "joint production, licensed production, sub-contracting in the purchasing country for components and spare parts, transfer of research and development capabilities, marketing rights, maintenance contracts for regional users of the weapon and imports of other industrial goods from the weapon recipient by the supplier country."[68] The amount of the offset in percentage terms varies, but even Saudi Arabia, a country with little manufacturing capability, demands that its multi-billion dollar packages with the United States and the United Kingdom be offset approximately 35 percent.[69]

This structural change was a result of several other changes already discussed. First, the demand for weapons in the 1980s continued despite the deteriorating financial conditions found in many developing countries. This was a function of continuing conflict and the need to provide for national security. DeCourcy's 1987 study shows that an overemphasis on the arms-for-oil phenomenon of the 1970s may have obscured the fact that both oil-rich *and* oil-poor states were demanding and acquiring arms.[70] The second structural condition that fostered the development of offsets was the rise of production capability in the Third World. Whatever the original motive for a developing country to develop or expand its defense industrial capability, once it existed, there was natural pressure to "develop a foreign market for its products as economic motivations overtake the original political and strategic ones."[71] These two conditions occurred just as the cost of producing military equipment accelerated. It has already been shown that one effect of this has been the slowing down of arms exports from these second-tier suppliers because of financial factors. But assuming the international demand is still there, offsets became an alternative means of financing increasingly expensive equipment in the face of declining reserves of hard currency. Given that the suppliers were also experiencing pressure to export in the face of defense budget declines and the prospect of their own defense industries going under, the presence of offsets as a mode of interaction is understandable as "an effort by the purchaser and the seller to make the best of unfavorable economic conditions"[72] in a world that continued to be ridden with conflict and the need for arms.

Whatever the form of the offset, it means that arms deals now include a wider variety of participants. This has led to a situation in which the government contracts for the military equipment, but the private companies involved in the deal must conclude and manage the offset arrangement. Offsets, therefore, have significantly increased the commercial nature of the arms trade. Additionally, offsets foster the trend toward trade in subsystems and components, since most recipients (including industrialized countries) have opted out of producing major end items. To the extent that offsets involve the transfer of technology and other facets of research and development potential, they have succeeded in accomplishing the goal of the recipient to enhance the capability to produce indigenous systems.[73] The most recent of three reports from the U.S. Congress Office of Technology Assessment, *The Global Arms Trade,* concludes as one of seven major findings that "cooperating with foreign industry in the development and production of weapons builds up their indigenous defense industrial capabilities, transferring potent, advanced defense technology to foreign nations."[74]

**From Bilateral Negotiations to Arms Bazaars.** Throughout the postwar and expansion systems, arms deals tended to be arranged on a bilateral basis. Even within alliances such as NATO, when multilateral organizational decisions were made to develop and produce a NATO weapons system, the actual payment and production arrangements were done on a bilateral, country-to-country basis. In the 1980-92 system, there was a definite shift away from bilateral national negotiations and dealings to the "arms bazaar" approach. Between September 21, 1987, and January 28, 1988, twenty-two military equipment exhibitions took place in eight different countries from Australia to Egypt. Just a few years earlier, only the Farnborough and Paris air shows were available.[75] *Jane's Defence Weekly* now publishes an annual *Defence & Aerospace Exhibitions Guide;* the 1991 edition (published December 22, 1990) was thirty-eight pages long and included information on eighty-six exhibitions. Attending these exhibitions is much less expensive and troublesome than traveling to a multitude of potential customer countries. For the smaller companies and brokers, the former approach was simply prohibitive and could be considered a deterrent to all but the major players in the arms trade. All of this reinforced some of the other structural changes depicted, especially the rise in importance of trade in subsystems on a more commercial basis.

**The Return of Illegal Arms Trade.** As previously mentioned when discussing the characteristics of the actors in the present system, genuinely illegal arms trade has reemerged after an absence going back to the pre-World War II era. The magnitude of this trade, and hence the extent to which its methods permeated the system, is a subject of much debate. In 1986, Michael Klare concluded that "black-market sales of $5 to $10 billion per year, coupled with a substantial trade in arms-making technology, could—if factored into the standard statistical tallies—eliminate much of the decline noted in military exports since 1983."[76] SIPRI's 1988 yearbook carried a treatment of the phenomenon for the first time, declaring that by 1987, "virtually all Western nations found themselves embroiled in controversies over illegal arms sales. Small black markets cater to North Korea, Libya, the IRA, and others. Much larger illegal opportunities arose when the U.S. led embargo on Iran forced the Tehran government to scavenge the planet in search of military hardware essential in its war with Iraq."[77] However, the 1990 SIPRI assessment of the illegal trade is that the "size of the market is impossible to ascertain, and so a true measure of its significance is also elusive."[78] A study of this illegal trade in the United States reveals a host of techniques of interaction (table 6–11) that, while not new to history, are new to the postwar era, not only in the United States but worldwide.[79]

**Gray Market Trade.** Along with this black market trade, significant trade arose in what is being called the gray market. As defined by Karp in the 1988 SIPRI yearbook, "the gray market includes officially approved exports from governments which do not want to be associated with their actions."[80] Some key examples given were the supply of Iran by North Korea and China, and the Israeli arms trade with South Africa; the infamous Iran-Contra affair in the United States is another example. As a *type* of trade, the gray market has always been a part of the cold war, but the discovery and public airing of so many cases seems to indicate that it has become more prevalent. As long as the international norm of government control holds, and the potential for negative consequences remains as part of the public's perception, gray market trade will continue as

Table 6–11
A Typology of Methods Employed in the Illegal Arms Trade

A. Acquisition of Equipment to Be Illegally Exported
    1. Theft
        a. Outright
        b. Manipulation of U.S. Military supply system
    2. Manufacture in U.S.
        a. Legal
        b. Illegal
    3. Legal domestic purchase by U.S. company
        a. Manufacturing firm
        b. Exporting firm
    4. Association with U.S. government
        a. Government contract
        b. Clearance held by employee
        c. Current or former member of U.S. military
    5. Surplus U.S. equipment

B. Fraudulent Licensing Procedures
    1. No license obtained
    2. Forgery
    3. Declaring item as low technology when it is high
    4. Declaring item for commerical use when it is military
    5. Fraudulent end-user certificate

C. Export Scheme
    1. Use third country as conduit
    2. Illegal redirection of shipment enroute
    3. Sell to foreign mission within United States
    4. Concealment of shipment
        a. Private ships and aircraft
        b. Carry-on luggage

governments who feel they must make a transfer in the name of national security may try to keep it private to avoid the political backlash.

**Dual-Use Equipment and Technology.** A final change was the increased trade in those items that have dual (both civilian and military) uses, such as computer chips, computers, and electronic systems that can be used in fighter aircraft as well as commercial aircraft. The presence of this market has always been a part of the postwar system. The COCOM agreement was signed in 1949 by the NATO allies and Japan in an attempt to prevent high technology from reaching the Soviet bloc and China. Dual-use equipment was of little concern as long as the United States produced all or most of the relevant technology to be controlled; however, as Europe recovered and Japan began to produce high-technology components and systems, controlling the trade became difficult. The first Reagan administration put a major emphasis on this issue, but when the East-West rationale disappeared in 1990, the natural commercial pressures to export commodities in demand saw this type of trade expand significantly, and the controls loosened.[81]

## Regime Norms and Rules

As mentioned previously, this system evolved during the 1980s, with its economic characteristics becoming more stable earlier than the political or military factors. But the abrupt changes that took place in the early 1990s provide a clear termination point for the system in 1992. As a result the regime rules which follow can be said to apply to the 1980-92 period as a whole, despite some of them emerging later in the system. In the conclusion of the book, alternative futures and their regime rules will be addressed.

### Conflict

**Major arms suppliers cannot control the outbreak of intrastate or interstate conflict using arms transfers as an instrument of policy.** Despite the fact that the cold war was still very much in evidence in the early 1980s, conventional warfare was thought of almost exclusively in terms of the developing world. Although the superpowers continued to supply their European bloc partners with those items at the high end of the technology spectrum to maintain the stalemate in Europe, none of these transactions were thought of as linkages between the arms trade system and the larger East-West struggle. Within the developing world, the proliferation of military capability and the general loosening of systemic political polarity have produced a situation in which the outbreak of conflict is, in essence, determined at the national and subnational level in specific geographic regions. There are minimal systemic pressures in place to deter one state or group from attempting to actualize its goals through the use of force. At the *intrastate* level, as in the previous period, the availability of weapons is such that there are few constraints on the commencement of this type of conflict. Once the underlying causes and more immediate reasons for most of the major insurgencies and civil wars of this period had surfaced, it was a foregone conclusion that armed violence would occur. There was little that could be done to stop the outbreak of political violence. The major *interstate* conflicts starting in this period (Iran-Iraq, Falklands, China-Vietnam, South Africa-Angola, Libya-Chad) also owe much of their existence to the proliferation of usable military capability that is the hallmark of this system. In this system, preconflict military balances are determined more by the recipient nations than by the suppliers. When any state feels that the benefits of starting a conflict outweigh the costs, it will commence using force to accomplish national goals.

In the aftermath of the Gulf War, there was no shortage of analysis that sought to put primary blame for the commencement of the war on the weapons that Iraq had acquired. Others sought to blame the United States for sending Saddam Hussein the message that an attack on Kuwait would not be opposed by the United States. In a sense, the outbreak of this war has not brought us much closer to resolving the dilemma of what causes wars—guns or people. It has, however, demonstrated that although the major suppliers could not affect the outbreak of war in this system, if the negative consequences are significant enough, it may be legitimate to ask once again the question regarding the role of armaments in the outbreak of war.

**The capability of the major arms suppliers to control the conduct and termination of intrastate and interstate conflict through the bilateral transfer of arms and other military capability is minimal.** Once armed conflict ensues, the larger question is how much control over the conduct and outcome of the conflict exists in the system. As previously mentioned, Neuman's study of wars occurring from 1977 through 1985 concludes that arms suppliers, particularly the superpowers, dictate the conduct and outcome of conflicts.[82] Her evidence is convincing, at least through 1985. For example, despite the U.S. attempts to convince Argentina that if it attacked the Falklands the full weight of U.S. arms transfer power would go to resupplying the British forces, it attacked anyway. Argentina had enough independent military capability to start the conflict, and given the various internal national dynamics, little could be done by the international system to stop it. The national forces of the United Kingdom also had enough independent capability to respond. However, once the armed conflict ensued, the outcome was heavily influenced by international systemic factors.[83]

Another case to examine is Afghanistan. In many respects, the Soviet invasion of 1979 was the result of the Soviet Union's inability to control the political situation through arms transfers. Once the USSR had taken direct control of the country, arms transfers were not a factor, except for the *mujahadeen* resistance. After a period of some reluctance, the United States began supplying those items of equipment, particularly the Stinger portable surface-to-air missile, that had a significant impact on the conduct of the war. There appears to be a consensus that this capability was instrumental in driving USSR military forces out of Afghanistan and having them revert to their former approach of arming their client state.[84] Arms transfers did not, however, succeed in shaping the termination of the conflict, as casualties continued to mount in a stalemate.

At the intrastate level, there were also some examples of how arms transfers can have an impact on the conduct and outcome of these lower-level conflicts. For example, in Sri Lanka, when the Tamil minority began being armed by Tamil nationalists in India, the magnitude of the conflict grew to the point where a negotiated political settlement became less likely.[85] As for the outcome of these types of conflict, the civil war in South Yemen in 1986 is a good example of how systemic arms supply factors come into play. In this conflict, both sides were armed with Soviet equipment, given that the USSR was the sole supplier to South Yemen. Despite the presence of Soviet advisors, little could be done to stop the conflict. As for the conduct of the armed conflict, Soviet advisors had to flee the country, and the fighting stopped after a few days when both sides had run out of ammunition.[86]

At the beginning of this discussion of the 1980–92 system, it was noted that the early 1980s represented a transition period in many respects. This is certainly true when looking at the effects of arms transfers on the conduct and outcome of interstate conflicts. As of 1985, despite some examples to the contrary, Neuman's conclusions about the influence of the superpowers were a valid description of the system. However, by the end of the decade it had become clear that conflict control through the use of arms supplies had changed significantly. The major reason was the diminution of the larger East-West conflict. Not only were recipient states now free to acquire large stocks of highly capable weapons systems, the suppliers were more free than ever to supply them without concern for the larger effects on the international political system. Put another way, the major suppliers could now say that regional conflicts were truly regional, with little potential for spilling over onto "home" territory.[87]

A brief look at the conflicts of 1990 makes this clear. For example, at the intrastate level, the Afghanistan war continued inconclusively as both the USSR and a conglomerate of Islamic and Western states continued to arm the respective combatants. This systemic change was also reflected in the Iran-Iraq war. By the end of 1986, the failed U.S. attempt to influence the outcome of the conflict had become public in the Iran-Contra hearings. The worldwide network of Iranian agents buying equipment on the black and gray markets to outflank the embargo by the United States had also become known. Not only were nonsuperpowers supplying a greater quantity of weapons systems, they were supplying those items that had became critical to the outcome of the conflict.

For Iraq, this took the form of French aircraft and Exocet missiles to attack Iranian shipping. Even the Scud-B missiles Iraq had received from the USSR had been upgraded indigenously (or, more accurately, through the importing on the open market of those technologies required to upgrade the surface-to-surface missile) so as to terrorize the civilian inhabitants of Tehran. In short, the system became one in which the control of the conduct and outcome of conflict through bilateral instrumental and structural influence attempts using arms transfers was minimal.

There were examples during this period when those charged with developing or overseeing the resolution of conflicts chose as their first and necessary step the disarming of the combatants; this was the case in Nicaragua, Lebanon,[88] Angola,[89] and Cambodia.[90] Operation Staunch, the U.S. effort to stem the flow of Western arms into Iran starting in 1986, could also be viewed as recognition that the level and sophistication of armaments affect conflict outcomes (however, the chances of success were minimal, given the preponderance of non-Western suppliers to Iran). Though peace may not break out upon the disarming of the combatants, there is little hope that any progress can be made while arms flow freely into the war zone. However, not enough of this type of activity has occurred to change the basic regime rule that once started, conflicts are hard to stop by controlling the flow of arms.

**For the first time in the postwar period, recipient states can now acquire enough military capability to deter, directly threaten, and influence the behavior of major states in the system.** The pressure to export even those technologies previously withheld for cold war reasons, combined with the proliferation of suppliers, created an apolitical arms transfer system that allowed many recipients to acquire enough usable military capability to have a direct impact on the major suppliers. The 1990 attack by Iraq on Kuwait and the subsequent U.S. response is not an isolated example. The proliferation of ballistic missiles and warheads of mass destruction had prompted the United States and other major actors in the system to respond in a variety of ways much earlier than the summer of 1990. The 1987 Missile Technology Control Regime, despite its clear nuclear dimension, is the first time that a serious international attempt had been made to control the export of conventional systems since the early 1950s in the Middle East. Even the USSR, not a signatory, later agreed to abide by its rules; this is not surprising, given that the emerging capabilities were in countries much closer to the USSR than other supplier states. World navies began to take seriously the buildup of the Indian navy. A veteran of the U.S. effort of the 1980s to stem the flow of high technology to the Soviet Union, Dennis Kloske, recently stated that "I think it is a brave new world that we are entering and it is clearly a reflection of the sobering prospect of having to face medium-size military powers that are armed to the teeth, that have developed their own very sizable and sophisticated industrial capabilities, such as Iraq."[91]

It is not the conclusion of this analysis that only arms are required to develop this capability to challenge major powers. As seen by the minimal performance of the Kuwait military forces in the recent war, twenty years of high-technology military imports were of little use. Some analysts suggest the same thing about Saudi Arabia. It also should be remembered that once engaged, the Iraqi military was soundly defeated by the coalition forces. However, without the East-West conflict to dictate national responses, Third World states have used arms imports to increase significantly their ability to raise the cost to major actors who wish to deter or dissuade them from those courses of action that may have negative consequences for the major actors in the system. It should be noted that this concern for the effects that Third World military capabilities may have on the supplier states themselves is truly unique to this system. Most previous attempts to control arms transfers that cited this danger were dismissed as unrealistic and not necessary.

*Foreign Policy and Diplomacy*

**Arms transfers are not a reliable method of securing political alignment and are decreasingly used by supplying states as a bargaining tool to gain strategic access or to influence the behavior and orientation of recipient states.** As long as the East-West competition existed as an overlay to the arms trade system, the use of arms transfers to secure short-term compliance in the behavior of clients was a regime rule. Even in the face of some abject failures, the superpowers continued to operate with this rationale dominating their export of arms. The changes of the late 1980s have basically eliminated this rationale from the system.[92] There is a great deal of evidence, however, that this has not been a sudden development but rather a gradual change throughout the 1980s. Take, for example, the policy statements of Secretary of State Haig regarding the purpose of arms transfers in the first Reagan administration. Arms transfers were to be used to create a "strategic consensus" against the real threat, the Soviet Union and its allies. This approach ran immediately into the reality of the Middle East. While pushing the sale of AWACS aircraft through the Congress in 1981, the Reagan administration attempted to portray a Saudi Arabia that wanted the AWACS to deter and defend against a Soviet attack on its Persian Gulf oil fields (the U.S. rationale for the sale). An exasperated Saudi oil minister Sheik Yamani, appearing at a UN meeting, finally made it clear that Israel, not the USSR, was the primary threat to Saudi Arabia. The arms trade system no longer demanded that Saudi Arabia, and other Middle Eastern recipients of U.S. arms, change their threat perceptions and behavior simply because they were receiving highly capable equipment from the United States.

Previously, recipient states in possession of strategically valuable real estate could exercise reverse influence. Negotiations for the continuance and acquisition of strategic access usually involved an arms transfer package. The decline in East-West conflict has seen even this aspect of the arms trade system begin to recede into the background, as the industrial states have less of a need for the access. Arms for access is still practiced[93] in those cases where base rights and access are still important, but the time is coming when strategic access will be gained more by commercial means (that is, rent) than having the recipient nation use reverse leverage to acquire modern armaments. It should also be noted that arms transfers can still have the important side effect of regional presence for the supplier, particularly if the deals require technical assistance and advisers. The fact that the United States had established a lengthy presence in the Persian Gulf through its arms transfers was invaluable when it came to deploying 500,000 troops to the region.

**Although some supplying states may continue to believe that arms transfers create structural influence, only in extreme cases does a structural arms transfer relationship guarantee that established supplier-recipient relationships and their resulting political and strategic alignments will be maintained.** One of the regime rules of the previous system was that, in reality, bargaining attempts using arms transfers rarely succeed. This may mean that the gradual decline in the 1980s of the use of this rationale (arms transfers as a bargaining instrument to secure short-term behavioral changes in clients) might have occurred as a result of gradual learning on the part of suppliers. But, as Krause points out, a better explanation is that the major suppliers continue to use the arms transfer instrument because they believe in the efficacy of structural influence.[94] In the real world of arms acquisition, much more is involved than simply acquiring end items. In most cases, there is a structure that is also acquired, including logistics, technical support, and (in many cases) doctrine. A recipient state that falls under such structural influence may not toe the line in the short term or in all cases, but generally will align itself with the political and strategic position of the supplier. India's alignment with her major arms supplier, the USSR, served as an example.

But even in the case of this structural influence, there was a serious decline in its prevalence as a regime characteristic. The urgency to export even the highest-level weapons systems and technology, as felt by the industrial-state suppliers in an age of declining procurement and research budgets, resulted in a situation where the cost of switching suppliers (that is, breaking loose from structural influence) became more than acceptable. Take the case of Iraq and the USSR. It is true that the USSR was the major supplier to Iraq during the latter's war with Iran. But the changes in the arms trade system allowed Iraq to break from this structural influence by gaining independence in those areas most influential in conducting the war. It the wake of that war, it has become clear that despite the continued acquisition of Soviet equipment, logistics, and even doctrine, Iraq did not feel bound to accede to the Soviet Union's request to halt it's war on Iran. It attacked Iran using Western-supplied systems.

Krause puts forth an interesting variant on this conclusion. He maintains that "although no supplier was able to exert *direct* influence over the fighting (and the achievement of a ceasefire), by forcing both combatants to seek arms from other sources and creating a costly stalemate the general goal of avoiding an Iranian military victory and eventually bringing about a ceasefire was achieved."[95] In this sense, he agrees with Neuman's conclusions regarding the influence of the superpowers in the Iran-Iraq war. However, this type of "influence"—letting the two combatants slug it out until they run out of steam—may be stretching the meaning of the influence concept a bit too much, not to mention its moral implications. In the policy world, particularly as it is observed when states justify their actions, citing the Iran-Iraq case as the result of structural influence is not purposeful enough to establish meaningful cause and effect. A much more accurate conclusion would be that in the 1980-92 system, increasingly devoid of the controls on arms transfers emanating from the cold war, political influence of all kinds using this instrument was rapidly disappearing.

**As the political visibility of the commodities being traded in the international arms market has declined, so has the utility of arms transfers as a political instrument.** In effect, what has happened is that arms transfers have begun to lose what can be termed their *political visibility*, a concept that can be fruitfully employed to explain the transformation of the system into its current configuration. The arms trade became not only more commercial but also more technical and, in fact, smaller in terms of the size of the commodities traded. There is a political difference between trade in fighter aircraft, tanks, and ships and the latest "black boxes" that double the detection range of the radar or enhance some other aspect of military capability. The effects of this decline in the political visibility of the commodities being traded can be seen throughout the system. Although the export of arms is outlawed in Japan, it quietly surged forth as a major supplier of those electronic components needed for advanced weapons systems. Germany has a restrictive national policy regarding arms exports, yet has great difficulty controlling the export of materials and technology related to conventional armaments, because of the small size of most firms and the difficulty in monitoring smaller-scale items. As pointed out earlier, the illegal market has returned as a result of this phenomenon. In the United States, a constant battle rages between the Commerce Department and the Defense Department that stems from the inability to agree on the political value of commodities. National governments continue to profess control at the national level, but the actual control has declined.

*Economic Effects*

**Despite a decline in the absolute number of defense production firms in the international system, a substantial number remain to meet the legitimate defense needs of a country or region. National and multinational firms and their supplier states are under significant economic pressure to export the maximum possible amount of their production.** By 1990, it had become clear that the trend of declining defense budgets that had been evolving since the mid-1980s was now a permanent reality and would continue in that direction. The major increases of the first Reagan administration were not matched by the other NATO nations. By 1986, the USSR, led by General Secretary Gorbachev, began what was to be a major campaign to wind down the military aspects of the cold war and to promote arms control. Despite foot-dragging within the USSR by 1990 major arms control agreements calling for the actual destruction of equipment had been signed and implemented. The nations of Europe agreed to reduce significantly the numbers of armored vehicles, artillery, and fighter aircraft on the continent, and to scrap those declared surplus by the agreement. All of this put the issue of the economic adjustment of defense departments, military forces, and their supporting industries on the front burner.

Every major industrial state in the arms production and trade system began to wrestle with three basic options, two of which directly involved the arms trade. The first option is to convert excess defense industries to the production of civilian goods. Many industries are finding this difficult, and many defense firms disappeared. But a significant number of firms will survive to meet legitimate national security needs in an uncertain and evolving international system. The second option, with which many firms are familiar because of previous downturns in domestic markets, is to increase the export of production to provide the additional revenue needed to survive given a smaller domestic requirement. The third option is for these firms to merge or integrate in some fashion. The result would be fewer (but stronger) arms production firms in a particular country, region, or industrial sector. No matter how much this will improve the economic condition of the resulting conglomerate, the small domestic market size in the industrialized supplier states still requires them to export a maximum amount of their production. One of the few pressures working against this option is the fact that many states who supplied both Kuwait and Iraq lost a great deal of money in the process. France is owed over $4 billion, Italy $1.5 billion, and the USSR (now Russia and the Ukraine) over 3.8 billion rubles. China lost a significant amount also.[96]

**Most of the recipients in the system cannot pay cash but still have a military or political need for acquiring advanced weapons systems. This leads to a system of creative financing, featuring offsets and national bank financing.** On the demand side of the equation, several regime rules appear to guide the behavior of recipients in the system. First, as has been well documented, most developing states are experiencing serious financial difficulties that appear certain to continue in the short to medium term. Given their debt problems, these countries are unable to pay cash for the type of weapons systems that they want for either legitimate military needs or prestige and political power. Credit has also become a problem, as international financial institutions are loathe to increase the indebtedness of a country that wants to spend the money on what is perceived to be the unnecessary acquisition of military equipment. But most recipients are viable enough economically that they can insist on offsets, credit from supplier country banks, and other creatively financed packages, given the dire straits in which most defense industrial firms find themselves.

**Arms industries in the developing world can produce for export only those low- to mid-level technology items that a smaller but more powerful group of firms in industrialized states are unwilling to produce.** The financial condition of most of the developing countries with arms industries has meant that they are unable to continue the progress they were experiencing in the early 1980s. Israel, Brazil, South Korea, and other leading arms producers in the developing world are all experiencing significant declines in their military-industrial sectors. This suggests that creative financing that includes offsets will succeed only on nonmilitary projects in the recipient country, or in terms of the coproduction of components for the acquired systems. Those few who could generate the capital (such as through another oil price boom) are not able to meet the infrastructure requirements. As a result, for all but a few countries, military production as a route to national industrialization is no longer viable, if it ever was. However, given the decline in the number of defense firms in the industrialized world and their concentration on the production of advanced and high-technology equipment, Third World industries can become suppliers of mid-level technology equipment, providing jobs and modest revenue.

**Exporting arms for the purpose of financial gain is a legitimate national objective.** During the 1980–92 period, the merchants-of-death taboo, which had been a regime rule since the 1930s, began to disappear. As the 1980s began, this rule existed at the nation-state level for all but the United States, the USSR, and China. The debates of the 1970s had reinforced the taboo against selling arms for money. But the Reagan administration reversed the anti–arms trade policy of the Carter administration by openly stating the importance of exporting arms to bolster the military-industrial complex. Later in the decade, the USSR began to rely more heavily on sales for economic benefits, and China appears to have completely abandoned its former emphasis on ideological and political criteria as it sells arms mainly for the money needed to develop its indigenous defense industries.[97] By the 1990s, few actors in the system were predisposed to forgo an arms sale for fear of evoking a negative merchants-of-death response. This is not to say that, at the national level, serious opposition to the arms trade was not motivated by this dimension. Rather, the systemic barrier to using this rationale had been seriously eroded.

*Arms Trade Control*

**The presence of armed states in the developing world that utilize imported advanced weapons systems to threaten the interests of the industrialized states has produced the international norm that trade in these type of systems can lead to negative consequences.** Under the regime rules of this system, it is possible for many developing states to acquire significant levels of usable military capability. We have also seen that controlling conflicts among these states, and preventing these states from threatening the interests of industrialized states, is difficult. Combined, these rules have resulted in these states utilizing this acquired capability to create negative consequences for the industrialized suppliers and for the system as a whole. As a result of such major wars as Iran-Iraq, the Falklands, and Iraq-Kuwait, all of which threatened the industrialized world in some way, a consensus among the major arms supplier states now exists that certain types of arms transfers can lead to negative consequences.[98] In the case of the proliferation of ballistic missiles with warheads of mass destruction, the consensus appears to have motivated the major suppliers and recipients to begin discussing methods to control and manage these negative consequences. By no means, however, has a consensus emerged regarding these methods.

**The negative consequences developing from the trade in advanced weapons systems are no longer susceptible to unilateral or bilateral control.** In a departure from the previous postwar systems, the decline of the East-West conflict basically eliminated most of the incentives for unilateral control exercised by the two superpowers as they both sought to control the use of their arms exports. In the 1980–92 system, they had both less interest and less capability to do so. Furthermore, no consensus existed on how to deal with the aforementioned negative consequences. In the words of a leading analyst of the international arms trade, "Will the current outrage about the West's prewar laissez faire policies toward Iraq lead to a reappraisal of arms export policies to other countries? Are the dangers of arms proliferation now so clear that countries will finally cooperate to establish an international system of restraint? The only answer is: It depends."[99] The problem is that within the international system, options other than arms control or restraint have a great deal of support.

Efforts are being made to deal with these negative consequences *after* the developing state has acquired the threatening capability, assuming that the "genie is out of the bottle" and that there is little that can be done in the way of controlling acquisition. Efforts in this category include the proliferation of tactical antiballistic missiles, which received a boost from the performance of the Patriot missile system in the Gulf War. It is just as likely that the industrial states threatened by these negative consequences will continue to rely on deterring or removing the threat militarily, especially if there is an international consensus such as that which existed when Iraq invaded Kuwait. A great deal of controversy surrounds the Gulf War on this point. Given Iraq's acquisition of military capability that could threaten the interests of the industrialized states, what did these states do to deter the invasion of Kuwait and other states in the region? If the conclusion is that the invasion of Kuwait by Iraq is an example of failed deterrence,[100] efforts at developing an arms restraint or control regime will receive less support.

**Multilateral arms trade control regimes will be most effective if they address negative consequences experienced by both recipient states in the developing world and industrialized arms-supplying states.** Although it is perhaps too early to be firm on this point, it appears that the control of interstate conflicts in the international system became more multilateral in nature in this period. The UN was very involved in the attempts to resolve conflicts in Angola, the Western Sahara, Afghanistan, Cambodia, the Persian Gulf, and Yugoslavia. If a consensus regarding the negative consequences of arms transfers holds, this may result in arms control regimes emerging to deal with these consequences. It is not clear, however, that a control regime focusing on exports (such as the Missile Technology Control Regime) will be the approach adopted by the system.

One feature of this system that remained unchanged is the resistance by those states requiring imported weapons systems to any multilateral arms trade regime that impinges on their ability to enhance their national security. Unlike the nuclear nonproliferation regime, conventional arms trade control regimes (for example, the MTCR) do not provide "carrots" as well as "sticks." One reason for the successful embargo of arms to Iraq after the invasion was the participation of arms recipients as well as suppliers. Egypt's proposal for arms trade restraint in the Middle East[101] may also indicate that recipient states recognize that there is an upper limit to their appetite for advanced weapons, beyond which the system responds to their definite disadvantage. In addition to all of the other lessons emerging from the case of Iraq, it also "is a salutary reminder of how a determined potential aggressor can circumvent international restrictions on the export of certain categories of weapons, if it really wants to."[102]

**Trade in conventional weapons systems such as fighter aircraft, tanks, and ships, along with their advanced missiles, continues to be an acceptable systemic norm. However, attempts by developing states to acquire ballistic missiles, especially with warheads of mass destruction, will be met with significant resistance through various unilateral and multilateral arms control mechanisms.** The effectiveness of arms transfer control regimes depends on the commodities being controlled. The action taken against Iraq in the wake of its invasion is a case in point. On one level, arms transfers increased as nervous states in the region sought to acquire more security in the face of an increasingly powerful Iraq. The arms included here were fighter aircraft, tanks, and other advanced but conventional weapon systems considered acceptable by the system. At both the G7 summit and the arms trade control talks in Paris by the five permanent members of the UN Security Council, care was taken to send the message that the supplier states had no intention of restricting those major end items considered necessary to defend sovereign territory. On the other hand, there is no question that the proliferation of ballistic missiles experienced a significant slowdown as a result of the consensus view that the system rejected those states that attempted to employ or threaten the use of the missiles. The

MTCR has been credited with deterring the development of the highly capable Condor II missile by Iraq, Argentina, and Egypt, discouraging the export of the M-9 surface-to-surface missile (SSM) to Syria by China, and forcing Germany to crack down on its illegal exports (which gave Libya a chemical weapons capability).[103] The MTCR's membership continues to grow. However, it failed to prevent Iraq from upgrading its Scud missile inventory and has many other well-agreed-upon weaknesses, especially the nonparticipation of China and North Korea.[104]

**Arms embargoes are a legitimate response to arms imports seen as threatening to the major industrial powers. They are more effective if they result from a universal condemnation of aggression and there is a consensus that further arms transfers to the aggressor state will lead to negative consequences to those conducting the embargo.** Arms embargoes serve as standard diplomatic tools in this system. This is a recognition that in addition to their military characteristics, weapons systems are seen as political commodities and a major symbol of political relationships. For example, in response to an airline bombing blamed on terrorists supported by Syria, the European Community embargoed all arms exports to Syria—an embargo which may soon be lifted to make another political point, rewarding Syria for its recent cooperation in the Middle East.[105] The European Community was also quick to ban arms sales to Yugoslavia during the latter's recent regional conflicts.[106]

The 1980–92 period saw the implementation of two arms embargoes with more military objectives. The first was Operation Staunch, implemented by the United States with the help of its NATO allies against Iran after 1983. Because of illegal arms transfer scandals occurring in most of the countries involved in the embargo attempt, the cooperation in restricting the flow to Iran was unusually high after 1986. One result was the shifting by Iran to non-European sources (see previous data). But the embargo did succeed in preventing the massive supply of spare parts for the U.S. equipment Iran needed to conduct truly effective military operations against the U.S. naval forces in the region. The second embargo, on Iraq after its invasion of Kuwait in August 1990, appears to have been very successful. In this case there was unanimity in the implementation, with it being very clear that breaking the embargo would have negative consequences for almost any nation involved.

**Unilateral and multilateral arms trade control measures designed to deter states from acquiring military capability though commercial channels, the black market, surplus equipment, and dual-use components and systems are very difficult to achieve.** Although there was some effective arms trade control in this system for advanced systems controlled by supplier governments, especially ballistic missiles, it should be remembered that several new trends have emerged that make controlling the trade more difficult. Arms transfers have become more commercial in nature. Should this develop to the point where there is a GATT-like regime—in other words, should arms begin to be treated more like tractors than weapons of destruction—controlling negative consequences will be very problematic. Additionally, this period saw the return of two types of transfers creating control problems: illegal trade, and the sudden availability of very lethal and effective surplus equipment as a result of arms control agreements and the decline in the defense budgets of the industrialized states. Finally, the increased trade in dual-purpose systems will be almost impossible to control. It is increasingly viewed as commercial and nonmilitary, despite the fact that such items as computers and their chips are indispensable to usable military power. It is the ultimate irony of this arms-trading system that, just when some modicum of universal cooperation in conflict control is emerging, the methods that can now be used to acquire the military capability to wage unacceptable armed conflict are much less susceptible to arms control measures.

Table 6–12 summarizes the 1980–92 arms trade system.

## Table 6–12
## Summary of 1980–92 Arms Trade System

| | |
|---|---|
| *Boundaries and Environment* | • 1980–1992<br>• Steady decline of bipolar political system means few arms deals based on east-west rivalries<br>• Continuance of regional conflicts insures constant demand for arms by LDCs<br>• Nationalism continues to dominate the international system, resulting in no formal international structure to control conflict or its causes<br>• Centrally controlled economies are discredited, and arms are exported more as commercial commodities<br>• Suppliers are mainly the national governments of major powers |
| *Characteristics of Units* | • Private national and multinational firms gain enough independence from national control to become major actors in the system<br>• Illegal arms traders enter the system<br>• Second-tier suppliers provide low- to mid-level technology armaments<br>• Recipients are primarily LDCs connected with regional conflicts and not the acquisition of prestige<br>• Recipients have serious lack of capital with which to buy arms |
| *Structure and Stratification* | • Overall decline in global arms trade to 1970s level<br>• Number of suppliers and recipients remain unchanged from previous system<br>• Majority of sales are to the developing world by the USSR, US and West European suppliers<br>• Buyers market created by declining domestic budgets in supplier countries<br>• Significant number of stable supplier-recipient relationships<br>• Conflicts in Middle East dominate demand<br>• Significant proliferation of usable advanced military capability to LDCs |

168 • *The International Arms Trade*

*Table 6-12 continued*

| | |
|---|---|
| Modes of Interaction | • Arms trade more commercial in nature<br>• Internationalization of production<br>• Offsets become primary mode of transfer<br>• Negotiating arms transfers becomes multilateral in scope<br>• Illegal arms trade returns to the system |
| *Regime Norms and Rules*<br>Conflict | • Major arms suppliers cannot control the outbreak of intrastate or interstate conflict using arms transfers as an instrument of policy.<br>• The capability of the major arms suppliers to control the conduct and termination of intrastate and interstate conflict through the bilateral transfer of arms and other military capability is minimal.<br>• For the first time in the postwar period, recipient states can now acquire enough military capability to deter, directly threaten, and influence the behavior of major states in the system. |
| Foreign Policy and Diplomacy | • Arms transfers are not a reliable method of securing political alignment and are decreasingly used by supplying states as a bargaining tool to gain strategic access or to influence the behavior and orientation of recipient states.<br>• Although some supplying states may continue to believe that arms transfers create structural influence, only in extreme cases does a structural arms transfer relationship guarantee that established supplier-recipient relationships and their resulting political and strategic alignments will be maintained.<br>• As the political visibility of the commodities being traded in the international arms market has declined, so has the utility of arms transfers as a political instrument. |
| Economic Effects | • Despite a decline in the absolute number of defense production firms in the international system, a substantial number remain to meet the legitimate defense needs of a country or region. National and multinational firms and their supplier states are under significant economic pressure to export the maximum possible amount of their production.<br>• Most of the recipients in the system cannot pay cash but still have a military or politicial need for acquiring advanced weapons systems. This leads to a system of creative financing featuring offsets and national bank financing.<br>• Arms industries in the developing world can produce for export only those low- to mid-level technology items that a smaller but more powerful group of firms in industrialized states are unwilling to produce. |
| Arms Trade Control | • The presence of armed states in the developing world that utilize imported advanced weapons systems to threaten the interests of the industrialized states has produced the international norm that trade in these type of systems can lead to negative consequences.<br>• The negative consequences developing from the trade in advanced weapons systems are no longer susceptible to unilateral or bilateral control.<br>• Multilateral arms trade control regimes will be most effective if they address negative consequences experienced by both recipient states in the developing world and industrialized arms-supplying states. |

*Table 6-12 continued*

- Trade in conventional weapons systems such as fighter aircraft, tanks, and ships, along with their advanced missiles, continues to be an acceptable systemic norm. Attempts by developing states to acquire ballistic missiles, especially with warheads of mass destruction, will be met with significant resistance through various unilateral and multilateral arms control mechanisms.
- Arms embargoes are a legitimate response to arms imports seen as threatening to the major industrial powers. They are more effective if they result from a universal condemnation of aggression and there is a consensus that further arms transfers to the aggressor state will lead to negative consequences to those conducting the embargo.
- Unilateral and multilateral arms trade control measures designed to deter states from acquiring military capability through commercial channels, the black market, surplus equipment, and dual-use components and systems are very difficult to achieve.

# 7
# Explaining National Arms Transfer Behavior and System Transformation at the Systemic Level

## Explaining Supplier and Recipient Behavior at the Systemic Level

The previous three chapters have defined four distinct international arms trade systems. No doubt, further investigation would reveal additional data that could lead to slightly different system characteristics and perhaps additional regime rules. And, as mentioned earlier, it is always difficult to maintain that fine line between putting forth strictly systemic data as opposed to defining systems in terms of national-level behavior. However, assuming that in the main the international arms trade has been adequately defined at the systemic level, and that these systems are roughly representative of the reality that is the arms trade world, a final question must be asked. How can these systems be used to explain and guide the behavior of suppliers and recipients within the system? This becomes the ultimate test of a system: its utility in explaining the behavior of its members and its transformation.

Why focus on the systemic level? As much of the research in international relations has shown, the information needed to explain the foreign policy behavior of states is often more readily available at the systemic level. This is especially true for an issue area such as the arms trade, where the behavior centers on a commodity that is wrapped in secrecy. At the systemic level, the data selected for analysis are produced by the entities called nation-states. These data tend to be observable and to be conceptualized using rules easily understood by most observers, particularly when compared with the other levels of analysis. The fact that the USSR supplied a client state with 100 tanks, and that the transaction was followed by the USSR using the client's port facilities, is the type of systemic-level information around which analysis can be conducted and consensus obtained. (This is not to say that measurement problems do not exist at this level; see chapter 2). Conversely, attributing motives to the USSR or the client state, or attempting to explain the action in terms of the bureaucratic politics of the Politburo, is in most cases more difficult. Even in the case of the open decision-making process in the United States, it is difficult to conduct analysis at this level.

A second and related benefit of systemic-level analysis is the currency of the data. Much of what we know about any given actor in an arms transfer system can eventually be obtained at the nation-state level. Memoirs, declassified data, and (in the case of scandals such as the Iran-Contra affair) data disclosed as a result of public demand all provide the stuff of which case studies are constructed that often unlock behavioral mysteries. However, this type of data may never be available, as in the case of closed systems such as China or the underdeveloped bureaucracies and personalistic systems found in the developing world. Be they scholars, diplomats, industrialists, or arms control negotiators, the current information researchers need to operate efficiently is often found only at the systemic level.

A third characteristic of systemic-level data is that consensus is easier to reach. Most of the debates about arms transfers center on varying interpretations of data found at the nation-state level. A good example is the question discussed in chapter 6 regarding the role of U.S. and USSR arms transfers in regional conflict. Neuman's work ascribes a major role to the superpowers, at least in those wars fought through the early 1980s. But her thesis became less convincing when applied to the Iran-Iraq war. Why? Whereas a great deal of nation-state level data was available for many of the earlier wars, the Iran-Iraq war was ongoing, and mainly systemic-level data were available. That data tended to point to the conclusion that the superpowers were in decline as arbiters of conflict outcomes: Iran was receiving most of its arms from China and North Korea, and Iraq had turned to Western Europe, particularly France. But the evidence needed to make firm conclusions regarding the effect of superpower arms transfers on the conflict remained at the nation-state level and for the most part unavailable to outside researchers. One might look at the behavior of the major actors and *assume* certain conditions and facts at the national level, but it is only the observables emanating from nation-states that can be used, at least in the near term. It is argued here that suppliers and recipients will more often draw lessons from systemic-level data rather than the competing explanations put forward by analysts with not only varying types of data but also various orientations and axes to grind.

Every major treatment of the arms trade in the literature has a section on rationales for both suppliers and recipients. Suppliers as well as recipients are said to have military, strategic, political, and economic reasons for supplying and acquiring arms. General agreement has been reached regarding a typology of these rationales,[1] which are then used to label certain countries as "restrictive" or "economic" suppliers, whereas certain recipients are labeled as acquiring arms strictly for prestige purposes. These styles of supply or acquisition are normally explained as being rooted in each nation-state's history and culture, and in effect, they become constants. The result is that when faced with the cognitive dissonance that occurs when countries do not behave according to their assigned styles, observers may treat the data as inaccurate, or the behavior as a one-time aberration or special case.

Why do supplier and recipient styles of behavior change over time? It is at this juncture in an analysis that a solid knowledge of the pressures on an actor from the system can produce a more accurate explanation. It is at this level that we can begin to conduct more accurately the most important function of political analysis; forecasting and prediction. Acceptance that systemic-level data are valuable in explaining the arms trade can be gained only after some consensus has been reached on the variables used and their conceptualization, as was discussed at length in chapter 2. This makes the task of projecting future systems easier, particularly when compared with the herculean task of forecasting the myriad variables needed in addressing suppliers and recipients at the nation-state level. Data at the nation-state level become indispensable at some point, but it is suggested here that they are more useful when they modify conclusions reached at the systemic level.

## Explaining Changes in Behavior across Temporal Systems

It is beyond the scope of this book to prove the utility of the systemic-level approach to explaining the international arms trade. That can only be determined empirically, based on how the systems put forth in this book are actually used by analysts and policymakers. However, to demonstrate the utility of the systemic approach and the specific systems defined in chapters 4, 5, and 6, a few illustrative examples can be given.

### The USSR as a Supplier

A good place to start is with the USSR, a supplier whose orientation and rationale for supply was perceived to be constant across all three postwar systems. The literature is unanimous in describing the Soviet Union as a state that supplied arms to extend its ideology and influence as it contested the United States for world leadership. But the actual Soviet arms transfer behavior in the postwar era changed over time and can be best explained at the systemic level. Even during the bipolar system, many of the Soviet arms transfers were directed less toward the United States than they were toward competition with its former ally China. But it is in the 1966–80 period of expansion that the systemic pressures on the USSR are more clear. For example, the USSR was not a participant in the dominant international economic system, yet it responded like all of the other major suppliers to the sudden rise in availability of petrodollars.[2] Analysts who focused on nation-state level analysis (for example, ideological and political rationales) were presented with an anomaly, especially when one realizes that two leading Soviet customers in this period were Libya and Iraq, two Islamic countries that had little use for communism. This cognitive dissonance was often dealt with by picturing the USSR style as "pragmatic," a label of little use in explaining the unknown or the future.

In the 1980–92 period, the changes within the USSR were an important element in explaining their arms transfer behavior. However, much of these changes were dictated by systemic shifts. Soviet hard currency earnings from arms sales peaked in 1977 and steadily dropped during the 1980s. Nothing currently known about the internal Soviet affairs of that period would explain this drop; throughout the 1980s, USSR military production remained as it had been from 1966 to 1980. As a final example of Soviet behavior, the USSR began to adjust to the free-enterprise nature of the current international system by adopting the sales techniques of the other major suppliers. Its representatives regularly attended the international weapons exhibits, aggressively displaying and marketing Soviet-made military goods. The suddenly commercial international arms market produced a Spanish company called East/West Engineering Company Limited. Its advertisement on the much-coveted centerfold of *Jane's Defence Weekly* in the fall of 1990 listed a full range of Soviet equipment, including T-72 tanks and SA-7 and SA-9 air defense missiles.[3] During the spring of 1991, the USSR was attempting to sell its latest combat aircraft to the Philippines, China, South Korea, and Israel, not a list of customers recognizable by anyone familiar with past Soviet exports.[4] In one final example, the USSR in 1991 offered its latest VSTOL aircraft, the YAK-141, to India. It had become clear to Soviet planners that without an export sale, it might not be economically feasible to produce this aircraft even for Soviet forces.[5]

None of this evidence is meant to detract from those excellent case studies conducted at the nation-state level. Rubenstein's assessment of Soviet arms transfers to Egypt provides insights and behavioral orientations that have obvious utility beyond that one case.[6] Carlson's research establishes that variations in the type, modernity, and quantity of Soviet naval exports were attributable to the bureaucratic political struggles taking place regarding the role of the Soviet navy.[7] Studies such as these clearly provide invaluable knowledge. However, the problem in using this type of analysis to generate conclusions regarding Soviet arms transfer behavior, as indicated above, is in establishing consensus on sources and data, as well as the availability of current data to assess current and future policy outcomes.

## The United States as an Arms Supplier

Much has been written about U.S. arms transfer policy and performance at the nation-state level of analysis, with little explicit treatment of the effects of the changing system on this behavior. Most studies of U.S. arms transfer policy focus exclusively on the great deal of information making up the foreign policy context, failing to integrate into the analysis the equally compelling systemic explanations for policy shifts. For example, the change from a Carter administration policy that attempted arms transfer restraint to a Reagan policy featuring arms export promotion has been the subject of a significant amount of analysis.[8] However, these studies tend to be couched in terms of factors such as the personal orientations of the presidents themselves, the competing conservative-liberal philosophies of the major political parties, or the condition of the U.S. industrial base. Few if any analysts note that this shift occurred just as the major systemic shifts described in this book were occurring. It is quite ironic that just at the time that U.S. defense industries had a president who was more supportive than any in the postwar period, systemic changes seriously affected their ability to sell arms abroad. The secretary of state even tried to resurrect the cold war as a rationale for arms transfers, despite the fact that most of the sales in the 1970s had been to states only on the periphery of the East-West struggle. Analysis at the nation-state level tends to put a premium on blaming someone or something within the nation-state. For example, in the 1970s it was the arms industries that took a lot of the blame when U.S. arms transfers were thought to be "out of control." Even at this level of analysis this argument ignores some very real facts, such as a Kissinger-led State Department that viewed arms sales as the primary instrument of foreign policy at the height of the post-Vietnam anti-intervention climate. But more importantly, explanations of U.S. arms transfer policy most often ignore systemic variables.

This is understandable for several reasons. Politicians see themselves as representing a nation-state that can change things to produce desirable outcomes; admitting that "the system made me do it" is a definite political liability. Secondly, explanations at the nation-state level (at least in the United States) are easier to see, even though a consensus is harder to reach. Often these reasons are more personal. They are also perceived to be easier to change, although this is not always true. For example, one of the major parts of the Carter restraint policy that the Reagan administration campaigned against was the emphasis on human rights. Several countries (such as Argentina) had been cut off from U.S. arms supplies because of their human rights abuses during the Carter administration. Yet the Reagan administration found it just as difficult to deal with this issue, and arms transfers to many countries continued to be restricted because of their human rights records. For example, only in December 1990 did the United States lift the arms embargo on Chile that was put in place by Congress in 1976, despite the fact that in 1981 Congress changed the law to allow the president to permit arms exports if he certified improvement in Chilean human rights behavior.

How did the United States respond to the systemic shifts occurring late in the 1980-92 period? First, "a distinctly economic component has entered U.S. international military sales policies in recent years," according to the congressional Office of Technology Assessment in its recent study.[9] Despite the fact that Secretary of Defense Cheney stated that he "would never be in a position of advocating arms sales simply for the sake of trying to generate business here at home," high-level U.S. Army and Air Force officers for the first time openly told Congress that the United States must export its production.[10] Additionally, the State Department urged all embassies to cooperate with defense industries as they market their products, a Center for Defense Trade has been created in the State Department, the U.S. ambassador to NATO has recommended that a GATT-like regime be established for defense trade, various proposals have surfaced calling for an export-import bank to support U.S. defense exports, and for the first time in a while, the U.S. government paid the bill for U.S. manufacturers to display their wares at the 1991 Paris Air Show.[11] These actions break with the actions of previous systems and can certainly be fruitfully explained using the characteristics and regime rules of the larger international arms trade system.

But these are policy choices that may or may not result in increasing U.S. arms exports. Just as these choices are dictated by systemic pressures, the actual policy outcomes are likely to be dictated by what is possible given systemic realities. For example, an increasing international interest in arms trade control has forced the United States to propose arms control measures as well as increased sales to the Middle East. The resulting cognitive dissonance[12] may very well see the public opt for fewer arms sales, particularly if the United States is seen as the largest arms exporter in a system increasingly concerned with the negative consequences of this trade. Initial reaction to executive-branch and congressional schemes regarding governmental financing for arms exports give us some clues in this regard. In the spring of 1991, a change to the Defense Production Act that allowed arms exports to be financed by the government's Export-Import Bank barely passed in the Senate. But it was defeated soundly in the House,[13] mainly because of the negative image of encouraging and enabling the sale of arms instead of the commodities normally supported by the bank.

A brief review of some basic U.S. arms transfer relationships with key recipient countries also reveals the effects of systemic pressures. For most of the postwar period, the United States did not waiver in its commitment to supplying Israel with advanced equipment. With the end of the cold war, though, and especially with the rising international concern with armaments levels in the Middle East, this relationship began to change. Israeli Defense Minister Moshe Arens was very outspoken about this change: "There isn't a single significant piece of equipment that is being sold only to Israel and not to any of the Arab armies. The commitment to assuring Israel a qualitative advantage is not being met and cannot be met as long as this is American policy."[14] Pakistan, long a U.S. ally in the cold war and the recipient of large quantities of U.S. arms, is now receiving no arms as a result of an embargo resulting from its nuclear weapons programs. Just a few years ago, despite a well-known effort to develop nuclear weapons, Pakistan was a favored customer because of the overriding interests of the United States in Afghanistan. A changed international political environment has clearly shaped this changed U.S.-Pakistani arms transfer relationship.[15] A few years ago, when the United States was interested in an opening to China, arms transfers to China became a foreign policy tool often used to align the two states further. As the cold war ends, however, the "China card" seems less relevant, and the United States has become more interested in post–cold war issues such as ballistic missile proliferation. Given the new international system, China's role as a major proliferator has resulted in its shift from friend to foe.[16]

*The U.S. Response to the Use of Offsets in Arms Deals*

A look at the specific U.S. response to several of the recent changes in the international arms trade system brings out the primacy of international systemic factors in explaining U.S. behavior. As noted in chapter 6, offsets have become a standard mode of transfer in the system. The United States did not invent this characteristic, and it can be said to be one of the last of the major suppliers to engage in this type of activity. Offsets became a reality as a result of political and economic inequities that drove recipients to seek ways to foster industrial development and to diversify their dependence in the international system. The data in a recent U.S. government report give a brief glimpse at how recipient countries vary in their offset requirements. When measuring offset obligations as a percentage of sales, the top five countries were Sweden (173 percent), Spain (132 percent), United Kingdom (105 percent), Belgium (86 percent), and Canada (78 percent). The lowest percentages of offsets were recorded by Israel (22 percent) and Singapore (30 percent).[17] Given this demand factor, the United States has responded as depicted in figure 7–1.

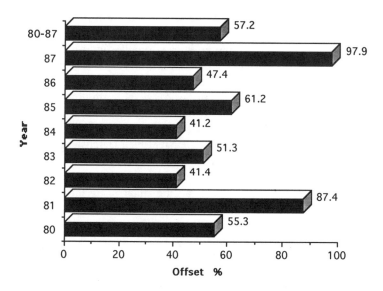

**Figure 7–1. Offsets as a Percentage of Contract Value, 1980–87**

Source: Office of Management and Budget Report to Congress, 12 January 1989.

An even more telling statistic of this growth is the number of employee years associated with military export sales contracts with offsets. In 1980, it was 3,400 employee years; by 1984, it was 14,800; and in 1987, it reached 26,300. There is no question that the trends above depict a United States responding to a buyers' market wherein the recipient countries of the world have adopted national offset policies. As long as this buyers' market holds, the recipient countries will continue to require offsets, and there is little that the United States can do about it.

The best evidence that the U.S. offset behavior in defense sales is *not* due to nation-state level forces is the amount of controversy these offsets created. The fiscal year 1990 defense authorization bill (PL 100-456) required the president, in effect, to submit an offsets-impact statement for each deal, evaluating the costs and benefits of offsets regarding the industrial base, employment, and national security itself.[18] This represents a Congress worried about domestic "buy American" pressures that are accelerating during a recession exacerbated by a decline in the U.S. defense budget. For its part, U.S. industry is determined to keep the U.S. government—both the executive and legislative branches—out of the picture so that they can respond to the international system that is, not the one that actors on the nation-state level would like to see. "It is our belief, and this is a belief held throughout the aerospace industry, that offsets are here to stay. Given this reality, it should remain within the purview of U.S. firms faced with offset demands to negotiate a realistic agreement in each instance.... Intervention by the U.S. Congress, no matter how well intended, will in all likelihood lead to undesirable results."[19]

## Illegal Arms Exports from the United States

Another change in the international arms trade system is the rise of illegal or black market trade. As many of the books on the "merchants of death" point out, there is no shortage of those who would use the plight of the insecure to make a profit by illegally selling arms.[20] However, if this greed is a constant in the international system, why was it that illegal trade surfaced in the United States only in the 1980s? The number of cases of illegal arms exports grew steadily during the 1980s, with several celebrated cases. In one instance, mentioned earlier, more than eighty Hughes 500D helicopters found their way to North Korea. In 1985 alone, there were 750 seizures worth $74.8 million in what most observers viewed as only the tip of the iceberg. Illegal trade of this magnitude did not exist in any of the previous postwar systems.[21]

Critics of the Reagan administration could find policies at the nation-state level that might explain this phenomenon, such as an arms export promotion policy, increased privatization in the defense industrial sector, and a munitions-control bureaucracy unprepared for the increase in commercial exports. However, it is at the international system level, where the context and regime rules are more supportive of illegal arms exports than at any time in the postwar period, that equally powerful explanations exist. First, the shift away from bilateral national negotiations and dealings to the arms-bazaar approach increases the overall number of contacts, which by necessity will include those looking for illegal exports and imports. Second, with offsets now a fact of life in making arms deals, arms deals must include a wider variety of participants, meaning that offsets have significantly increased the role of the private sector in arms trade. This privatization of the trade does not equate to illegality; however, given a normal distribution of criminal intent throughout the United States and the world, an increase in illegal sales should not come as a surprise. Third, the increasing percentage of trade in spare parts and dual-use equipment results in an overall decline in the visibility of arms exports and increases the opportunity for illegality. Fourth, a particularly important change in the structure is the legitimation of private arms dealers. The Iran-Contra revelations were but the latest piece of evidence that private arms dealers and middlemen are increasingly being used by a wide variety of governments in the so-called gray market.[22] Such activity is "legal," since it is controlled by governments most of the time. An increasing use of private arms dealers by governments may have produced some positive policy outcomes, but not without the side effects of legitimizing their place in the system and diminishing the capability of governments to control the trade.

A fifth structural change, the growing stockpile of surplus and uncontrolled equipment, simply makes it much easier for those engaged in illegal trade to acquire commodities to trade. Sixth, there is now the presence of a supply network developed for the drug trade that is increasingly being used for arms. Perhaps one of the most critical differences between the 1980s and previous periods, this supply network is efficient, capable of handling large commodities of goods (legal and illegal), and clandestine. The revelations in recent years of the methods used by international drug dealers make it clear that this network is being used for arms as well as drugs.[23] Even with the end of the Iran-Iraq war, one of the major conflicts producing demand for illegal arms, this network is still in place and can facilitate the flow of arms into other conflicts where black market purchases would be preferable to those from governments that might try to apply political leverage.

In addition to these structural changes, it should be noted that the 1980s saw a concentration of conflicts that created unusually strong pressures for the illegal acquisition of arms from the United States. Iran's war with Iraq is one obvious example. Iraq, due to the disparity between its population and Iran's, was forced to adopt a strategy that involved maximum use of its airpower to attack Iranian oil facilities. Iran's air force and air defense missile capability were U.S.-made, thus creating the pressure to establish an all-out Iranian effort to obtain the needed spare parts from the United States. Open warfare between Iran and Iraq has ceased, but there is no shortage of international phenomena creating demand for illegal exports from the United States. A good example is the proliferation of medium-range surface-to-surface missiles with conventional warheads, as discussed in chapter 6. The United States signaled its concern over this development by agreeing with the European allies to restrict the flow of these missiles and any technology and equipment that would foster their manufacture. However, the United States was not immune from the pressure on developing states to acquire this capability illegally. On June 24, 1988, two Egyptian military officers and three Americans were charged with plotting to export illegally a box of carbon-carbon on an Egyptian military transport leaving from Baltimore. This sensitive material is tightly controlled by the United States as it is used in making warheads more accurate. U.S. officials claimed that Egypt planned to use the material in a joint program with Argentina to develop the Condor II surface-to-surface missile with Iraqi funds.[24]

Changes in the structure of the international arms trade are not direct causes of the illegal arms trade but rather conditional or intervening variables in the equation. The demand for illegal arms created by Iran's war with Iraq, for example, was a more direct explanation. But because the new structure allowed Iranian agents (and those desirous of ballistic missile technology) to make contacts more easily, and because equipment is harder to trace given the realities of offsets, internationalization, spare parts, and high technology in small packages, the structure has been instrumental in complicating efforts to control the illegal trade.

## Second-Tier Suppliers

Another fruitful source of examples is the group of suppliers who entered the system after its characteristics and rules had been established prior to World War II. This would be the case for most of the second-tier suppliers, but Brazil is a good example. It is true that the rise of Brazilian defense industries was partly a function of national goals, budget decisions, and the particular methods chosen to link civilian industry with the military in Brazil. However, a look at the actual export of Brazilian arms reveals a record significantly influenced by systemic changes. In the 1970s, oil-poor Brazil found a natural outlet for its fledgling arms industries in Middle Eastern countries with a surplus of petrodollars. Major sales were made to Libya, then Iraq, allowing Brazil to use this capital to develop further systems that could compete in the international market. This was particularly true for its aircraft, with Brazil eventually selling a military trainer aircraft to the United Kingdom. However, with the onset of the debt crisis, the overall decline in the ability of developing countries to pay cash for arms, the end of the Iran-Iraq war, and the arms embargo on Iraq in 1990, Brazil's sales plummeted.[25] As of 1991, its defense industries were in a crisis, with several major firms having gone bankrupt. Brazilian arms sales are almost perfectly correlated with world demand, and nothing within Brazil explains this plunge, except for the fact that the larger powers were better equipped to deal with the lack of recipient cash because of their larger domestic markets.[26]

Argentina and Israel have experienced similar fates to Brazil's.[27] Turkey, meanwhile, presents an interesting response to systemic change. For the past few years, Turkey has been engaged in a major effort to develop its defense industries. It has done so using offsets, particularly with the United States. For example, General Dynamics won a major contract to sell F-16s to Turkey, but only after agreeing to develop an aircraft industry in Turkey to coproduce the aircraft. A similar deal was struck with the FMC Corporation for the coproduction of armored infantry fighting vehicles. However, as a result of the surplus military equipment generated by the CFE agreements in Europe—fairly modern equipment, to be transferred at very low cost—Turkey is rethinking the option of producing its own equipment, especially given that the latter is extremely expensive in comparison. This rethinking is no doubt spurred on by the shrinking international market for arms, in which Turkey would not be very competitive as an exporter.[28]

Another example is that of China, a major arms supplier in the 1945-60 period that dropped out of the business, only to return to a system quite different than the one it left. In the immediate postwar period, particularly after its split with the USSR in 1959, China was one of the world's major suppliers and drew quite a bit of attention as one of North Vietnam's major suppliers. Its annual exports from 1958 to 1960 were $505 million, $439 million, and $406 million, respectively, whereas its annual average for the 1961-70 period was $92 million. Beginning in 1971, China's exports once again became significant for a second-tier supplier. Its recipients in both the 1945-60 and 1966-80 periods were mainly poor states being wooed for their opposition to the USSR.[29] However, China did begin to take advantage of the system, seeking to become more independent in the production of advanced military equipment. Its exports to Egypt and Pakistan, commencing in the 1970s, signaled a response designed to generate cash, since both of these states were recipients of petrodollar largess from the Arab oil states.

In the current system, there is a consensus that China has become a commercial supplier willing to sell most anything in order to generate capital for indigenous programs.[30] The most recent CRS data, published in August 1991, shows China ranked as the number three arms supplier in the world for 1990, with more than $2.6 billion in orders—just ahead of France, which had $2.3 billion.[31] Like the former Soviet Union, China is very aggressive in advertising its military goods, and unlike the USSR, sponsors several annual military exhibits to foster such activity. Tiananmen Square has demonstrated that the much-predicted internal reforms have yet to take hold within China, but its arms transfer behavior depicts an actor very much responding to the loosening of the international arms trade system.[32]

One of the leading suppliers in the current system, at least until 1989, was Czechoslovakia. One of its major roles in Eastern bloc's Council for Mutual Economic Assistance (CMEA) and the Warsaw Pact was to produce and export massive quantities of major defense equipment, particularly trainer aircraft and tanks. In what is probably one of the purest examples of systemic shifts affecting national behavior, the end of the cold war has resulted in its disappearance as a major arms supplier. After an initial declaration that Czechoslovakia would cease the "immoral" activity of arms exports,[33] internal economic pressures have kept some of the tank factories alive for the moment.[34] But it is clear that the system will not allow or tolerate such exports for much longer.

The overall behavior and performance of the second-tier suppliers can be usefully explained as a function of the shifts in the international system. Just as they were being perceived as potential competitors to the major powers in the early 1980s, even in some areas of advanced technology, major systemic shifts saw them quickly revert back to the role they now hold. The cost of advanced technology and the overall decline in the economic status of this level of supplier, both systemic factors, combined to doom them to that of niche suppliers for those low- to mid-level technologies needed by states who actually intend to use equipment in armed conflict and cannot or do not want to acquire such systems from the major industrialized suppliers. Even this niche, however, is not secure. Although it was once thought that second-tier suppliers would serve as a way out for recipients who sought to diversify their dependence, the sudden collapse of the East-West conflict makes the whole concept of dependence and influence much less salient. Major supplier states may now seek to reenter these niche markets.

## The Response of Recipients to Changes in Arms Transfer Systems

As with the USSR as a supplier, it would be useful to select a recipient whose basic foreign policy determinants and context have remained basically constant throughout the postwar period and show how its acquisition behavior has varied. One such example is Egypt. In the 1945–66 period, Egypt went from being a monarchy greatly influenced by its former protector (the United Kingdom) to becoming an independent and nationalistic Arab state that vied for leadership of the Third World. Most of its military equipment was British (with some French and American) until 1955, when Nasser, thwarted in his quest to challenge the developed world, switched suppliers to the Eastern bloc. The massive arms transfers to Egypt from Czechoslovakia in 1955 were a result of the interaction of this recipient demand and the bipolar arms trade system, which provided only one other source of supply. As Egypt continued a foreign policy designed to lead the Arab world against Israel, Egypt found itself on the losing end of armed conflicts in 1956 and 1967. The arms transfer regime rules of that period help explain these losses, particularly the extensive control that the supplier, the USSR, had over the use of the arms. At the end of this period, Egypt—despite a radical regime dedicated to a set of foreign policy goals—required a military capability that the system would not allow.

In the 1966–80 period, Egypt's unchanged foreign policy resulted in a very different arms acquisition pattern, in part attributable to the change in the international arms trade system. For example, Egypt broke off its arms transfer relationship with the USSR in the wake of the 1973 war with Israel. Although Egypt had been able to score some initial successes without the USSR (its advisors had been removed a month before the attack), in the end it was the superpowers who dictated the outcome (see the regime rules for this period in chapter 5). In an attempt to further its foreign policy goals, Egypt broke off its arms transfer relationship with the USSR and began to move into the Western camp. The international arms trade system was such that only the United States could supply Egypt with the modern arms needed for future conflicts in the Middle East. This meant that, despite a bold foreign policy decision that had unanimous national support, it was many years before the international system would allow Egypt to switch suppliers successfully. In 1978, Egypt and Israel signed the Camp David accords, but not before Egypt ensured that the United States would supply massive amounts of modern military equipment. This was a sign that Egypt was well aware of its military needs, and that only the United States was in a position to satisfy them.

During the 1980s, Egypt and its foreign policy of increasing its independence again responded to the changing international system by positioning itself as a second-tier arms supplier. It is interesting to note that during the early 1980s, when its main supplier, the United States, was attempting to couch its arms transfers in East-West terms (the strategic-consensus concept), Egypt responded with rhetoric and rationales that spoke of its hostile neighbors (for example, Libya) as clients of the Soviet Union. Egypt also used an international system whose modes of transfer now were more commercial and involved industrial offsets to build up a defense industry large enough to have a major impact on the Iran-Iraq war. It could be said that Egypt, which had been ostracized in the Arab world as a result of Camp David, worked its way back into the good graces of its Arab brethren through the use of the arms transfers to Iraq. These transfers included indigenously produced equipment, as well as a significant amount of Soviet equipment now surplus as a result of Egypt's switch in suppliers. By 1990, Egypt was coproducing a host of weapons systems with the United States, several Western European states, and second-tier suppliers such as Brazil. Its armed forces participated in the action against Iraq, not as a compliant ally of the United States but as an independent actor in the region. What had changed was not so much the foreign policy goals of Egypt, but rather an international system that now allowed arms acquisitions from many sources and the elimination of the use of arms by major powers to influence the behavior of client states.

*The Arming of Iraq: A Systemic Explanation*

When Iraq attacked Kuwait in August 1990, most of the world asked why. How could a country in poor financial condition, recovering from a very costly war with Iran that had only ended in 1988, acquire such a modern military machine, conquer a neighbor state, and hold a massive military force from the developed world at bay? Though many explanations for the attack emerged, it became clear that a major factor leading to the attack itself was the newfound confidence gained by the acquisition of destabilizing weapons systems. A review of the press coverage, as well as of the rhetoric used by world leaders opposed to the Iraqi acquisition and use of these systems, tended to seek explanations at the personal and nation-state level. Rarely was this situation addressed in public without using the name of Saddam Hussein. Biographies of this man quickly surfaced, and the world soon knew what only area experts (and the people of Iraq and its neighbors) had long known. Here was a man who had been dedicated for over twenty years to leading Iraq into a position of dominance in the region, a dominance that would be based mainly on military power. It surprised few of these experts that he was apparently willing to use chemical weapons against Israel, for example.

So explanations at the personal level were of little utility, since Saddam Hussein had not changed. What of factors at the nation-state level? Again, it can be seen that for many years Iraq's political system had become increasingly authoritarian and personalistic, so that there was no opposition to whatever was desired by Hussein. The foreign policy goals he set out were Iraq's goals. Additionally, the military-industrial complex, the structure of the armed forces, the desired weapons systems to be acquired, and the doctrine to be used in military operations were planned many years ago. What we are left with is the reality that most of the explanation for the acquisition and use of these weapons systems lies at the international system level, and particularly within the arms trade system.

By the late 1980s, the larger international system had shifted away from the East-West bipolar system of the past forty years. Iraq therefore took advantage of a world that was breathing a sigh of relief that the threat of East-West conflict was diminishing. It diversified its suppliers and took advantage of the fact that so many new suppliers had entered the system. For example, many of its armored personnel carriers and some rockets were purchased from Brazil, and long-range guns were purchased from South Africa. The rise of strictly commercial trade in arms and particularly dual-use items allowed Iraq to use its oil money (actually, Saudi and Kuwaiti oil money) to buy directly from Western firms who found their governments somewhat looser in controlling exports, especially as defense budgets in developed countries began to decline late in the decade. When focusing on a specific weapons system—for example, the land-based surface-to-surface missile—the power of systemic-level factors in explaining Iraqi acquisition behavior is striking. In 1974–75, with an international arms trade system with a bipolar flavor and trade in off-the-shelf end items as the dominant mode of transfer, the USSR supplied Iraq with the basic Scud-B missile. Egypt, North Korea, Libya, Syria, and South Yemen also received this mobile SSM, which had a range of 280 kilometers. At that time, the missile was difficult for these countries to use, and the Soviets believed that its use would be "controlled," which also fit with the basic regime rule that the superpowers retained the capability to influence conflict outcomes. But in the 1980s, Iraq began to acquire the technical assistance and expertise needed to upgrade this basic Scud system to develop the al-Hussein SSM, which has a range of 900 kilometers and was used extensively against Iran. Later in the 1980s, the Iraqis added the capability to launch chemical warheads using this SSM and, in fact, threatened to do just that to Israel in the spring of 1990.[35] Iraqi foreign policy goals and the determination of the military to deal effectively with Israel and Iran had changed little since the early 1970s. What had changed was a system that now allowed Iraq to develop and acquire an SSM that at last gave it a credible threat to back up its foreign policy goals.

Why the continued emphasis on personal and nation-state level analysis when explaining Iraq's arms acquisition behavior? As mentioned earlier, in an international system where the use of diplomacy and force remains a national response, however it may be coordinated, it will be national leaders who are charged with coming up with the rationales for the use of the various instruments of policy. Until the day when there is a truly international military force, explanations for aggression, poverty, and all of the other problems of the world will be shaped by the aggregation of national perceptions. Leaders of nation-states are charged by their citizens to "do something," either against or for another nation-state. This usually results in assessments that blame national actors, not the system. Two recent examples make the point. In December 1990, the *Sunday Times* "Insight" team in London published an exposé of how British industrialists, coached by an assistant minister of defense, circumvented the arms embargo of Iraq and supplied it with advanced technology used to invade Kuwait and to challenge the British troops deployed in Saudi Arabia.[36] At almost the same time, the *Wall Street Journal* published a similar story about U.S. firms supplying Iraq with similar capability.[37] When the reader is finished, there is someone to blame and the solutions are national, not international, in scope.

There were explanations that at first glance appeared to be at the international systemic level. For example, shortly after the invasion of Kuwait in August 1990, the Public Broadcasting System in the United States aired one of its "Frontline" presentations, entitled "The Arming of Iraq." In the first minute of the broadcast, the narrator stated that the massive military response to the invasion stood as evidence that "the *world* has made a big mistake in failing to control the arsenal of Saddam Hussein"(emphasis added)[38] The show was divided into six segments—the chemicals, the missiles, the bank, the dealer, the watchdogs, and the policy (of the United States). The broadcast provided myriad details on how the international arms trade system works, but in the end, the causes and policy remedies remained at the national level. This is particularly true of the last segment, which ended with a former high-level U.S. official feeling some remorse that "we" did not succeed, that "we" didn't get the outcome we wanted. The narrator concluded the broadcast by citing the most important lesson to be learned—that "it is a mistake to ignore principle in the name of *realpolitik*. It is a mistake to support those who share neither our values or our goals. It is a lesson we repeatedly forget." Although many will argue that the U.S. policy in this case was flawed, preventing the future occurrence of negative consequences from arms exports is hardly possible only at the national level. Not once in this broadcast, after the first mention of the word *world,* did the narrator allude to the arms trade as a systemic reality.

Had systemic-level explanations been put forward by national leaders, this would have shifted the emphasis toward systemic-level solutions. In the case of Iraq, had all of the major powers that supplied Iraq with the wherewithal to attack Kuwait confessed and admitted that it was the system that was at fault, it would mean that some sort of a multinational arms trade control regime would have to be promulgated to prevent this from happening again in Iraq and elsewhere. At a minimum, a systemic explanation for Iraq's behavior would shift the effort to control the negative consequences of arms transfers to a different set of variables. As this book has shown, in the aftermath of the invasion of Kuwait such a systemic effort is far from being a reality, despite the fact that the five permanent members of the UN Security Council met in July and October of 1991 and in February 1992 to discuss arms trade control measures. The U.N. resolution of December 1991 establishing an arms trade register has been turned over to a panel from eighteen countries for the development of implementation procedures. While such a confidence-building measure has clear antecedents in the Persian Gulf war, compliance in the form of voluntary reporting of arms exports and imports by U.N. member states will not be known for several years. It will take that long to accomplish its main goal, providing early warning of any buildup of destabilizing weapons inventories.

This necessarily brief recounting of the Iraqi acquisition of advanced weapons systems is designed to reveal a bias toward national- and personal-level explanations, explanations that are always controversial because of a lack of the current and complete information required to reach a consensus. But the systemic-level explanation is both current and less contentious, and it reinforces the utility of explaining foreign policy outcomes at this level of analysis.

## *Explaining the Behavior of Subnational Actors: The Case of the U.S. Defense Industries*

The characteristics and regime rules of the international arms trade system can also be used to explain the behavior of subnational actors such as presidents, legislatures, and a host of domestic actors who make up the mosaic that is the foreign policy-making context. As one example, the defense industry in the United States can be seen responding to the shifts in the international arms trade systems.

As a result of the various conferences and efforts by the League of Nations in the 1930s, most countries did agree to exert *national* control of their armaments industries, including exports. It was at this time, for example, that the United States first developed the munitions-control list, which is alive and well even today. Also developed in this period was the idea of an end-user certificate that the recipient country signed, agreeing that the defense products being received could not be transferred. In short, the idea that "arms are not refrigerators" was codified on a national and international basis. It should be remembered that the cause of all of these changes was the charge that those manufacturing military equipment had become merchants of death. This is important to remember, because this syndrome has not died and haunts U.S. industry at every turn. One of the implications of the changes of the 1930s was that the *moral* aspect of exporting arms had shifted from industry to government. Charges of "merchants of death" could no longer be hurled at an industry whose exports were controlled by the government. As can be seen in the excerpts from congressional testimony, today's defense industrialists are quick to point this out to critics of their activities.

In the postwar period, very little controversy surrounded the role of U.S. industry in the arms trade. For one thing, a significant percentage of arms exported from the United States were in the form of grant aid of surplus equipment. Industry was busy developing new weapons for the U.S. defense buildup against the USSR. Secondly, the idea of arming others was not very controversial, given the cold war and the domestic consensus that a military response was appropriate. This is borne out when looking at the lack of testimony by industry representatives at the various hearings held on security assistance. It should be noted that this is in stark contrast to the overall lobbying effort pursued by industry in regard to the U.S. defense budget, which has always been significant.

The defense industry suffered serious setbacks during the late 1960s, as United States involvement in Vietnam declined and major cuts were made in the procurement and research and development sectors of the U.S. defense budget. Faced with a domestic consensus for this decline, industry had several alternatives. The first was to get out of the business, which many did; by 1979, the number of U.S. firms in the defense business had declined from 6,000 in 1968 to around 2,000. A second alternative was to hang on, lobby for increased expenditures, and wait for the eventual upswing in defense production. This is what happened, as the collapse in détente, the failure to build on arms control agreements, and the increase in military intervention by the Soviet Union (culminating with the invasion of Afghanistan in December 1979) interacted to cause the defense budget to increase by the late 1970s. Ronald Reagan was elected with an overwhelming consensus to increase defense spending and to do something about the ailing defense industrial sector.

But in the late 1960s, many companies selected a third option, responding to their business instincts by searching for non-U.S. markets. Several factors aided in this effort. First, U.S. security assistance had been gradually shifting away from grant aid, and by this time a preponderance of arms exports were foreign military sales, either cash or credits. Secondly, the "Nixon Doctrine" specifically called upon the United States to honor its international commitments not through Vietnam-style intervention but rather with arms exports and advisers. Secretary of State Kissinger became known for his frequent use of this instrument of foreign policy to garner diplomatic gains. Third, for the first time in the postwar period, the United States was experiencing a trade deficit, and Nixon responded with an export promotion policy, easing restrictions on commercial firms—including those in the defense sector. The onset of détente aided in this development, as fears of technology giveaways to the Soviet bloc were minimized. A fifth development was the growing feeling in the Congress, generated mainly by anti-Vietnam sentiment, that the government should be getting out of the arms sales business. Senator Fulbright actually proposed legislation to this effect in December 1973:

> In order to reduce the role of the United States Government in the furnishing of defense articles and defense services to foreign countries and international organizations, and return such transactions to commercial channels, the United States Government shall reduce its sales, credit sales, and guaranties of such articles and defense services as soon as, and to the maximum extent, practicable.[39]

A sixth, and perhaps most important, development was the huge increase in money available for arms purchases in developing countries following the oil price increases by OPEC. The result was a significant increase in U.S. arms exports and a relative decline in those arms export controls that so irritated industry. Simply put, industry had no reason to testify or otherwise complain about a very positive picture from their perspective.

By the mid-1970s, two trends intersected to arouse the defense industry in regard to the export of arms. First, the West Europeans had recovered from World War II and were serious competitors, particularly in the Third World. Despite record overseas sales, industry could see serious market-share problems down the road. Second, Senator Fulbright's proposed amendment aroused not only Congress but also the executive branch. Both the Defense and State Departments testified that they would not welcome this very important diplomatic tool being relegated to the commercial sector. As a result, a series of legislative acts were taken, eventually culminating in the Nelson Amendment in 1974, which gave Congress a larger say in arms sales decisions. It is at this point that we begin to see industry representatives speaking out, first against the proposed changes in legislation by an aroused Congress, and then against the arms transfer restraint policy of the Carter administration. Despite the election of President Reagan, whose policies were extremely supportive of arms exports, the shifts in the international arms trade markets and concomitant decline in the U.S. share of the market have seen industry continue to be active in representing its views as it strives to increase exports.

Each year since 1977, an industry spokesman has presented the general view of industry toward current or proposed policy actions by the government before the House Foreign Affairs and Senate Foreign Relations Committees as they hold hearings on the foreign and security assistance bills. Most of the testimony is from an organization called the American League for Exports and Security Assistance (ALESA). ALESA was first known as the American League for International Security Assistance and was created in the 1970s for the express purpose of dealing with developments outlined above.

Many themes emerge from a reading of industry testimony at congressional hearings during this period. One is that government may be very involved in controlling exports or even making the arrangements, but in the end, it is industry that must execute the contract. Industrialists call for letting "industry take it from there." Secondly, industry fully understands the role of government —that arms are not refrigerators, but rather an important instrument of foreign and national security policy. All that is desired is certainty regarding the rules and policies being developed. A third and quite natural theme is that industry is most worried about the economic effects of excessive controls. Jobs, the balance of payments, and the survival of industries are paramount. A fourth theme is, simply, fewer restrictions. This comes out clearly in that testimony where industry put forth evidence as to business lost during the Carter administration as a result of restrictions. In short, industry responded to the 1966–80 international system by first warding off governmental interference as it attempted to take advantage of the petrodollar explosion, and later trying to clarify rules within which its members could act as normal commercial firms.

In the 1980s, industry was very satisfied with the Reagan approach to arms sales but was faced with a series of developments requiring renewed efforts on their part to shape policy and laws. The first of these was the increased use of foreign military sales credits because of the decline in money available in the Third World. Industry was trying to ensure that these credits would not be used to develop future competitors in other countries. Secondly, as offsets became the norm, Congress became very active in trying to stem their increase and the subsequent loss of U.S. jobs to other countries. The testimony of a Boeing official, cited previously (see note 19 for this chapter), reconfirms the industry orientation toward minimal government interference as the method most likely to produce maximum benefits to the United States as a whole.

It is true that U.S. defense industries, and especially their representatives in Washington, must first pay attention to the national-level forces, such as the orientations of key individuals and organizations that must pass on their requests for arms exports. They also pay a great deal of attention to the rules and process itself. But in the final analysis, these national-level forces have changed very little, especially since the 1966–80 system saw grant aid come to a close and more commercial modes of transfer ascend in importance. It remains important for industry to clarify and certify the national "rules of the road." But once done, and it is argued here that this is not a contentious issue, what remains as an explanation of how industries behave is found at the international systemic level.

In the case of offsets, for example, what explains the fact that any industry wishing to compete for international sales must have an office that handles the offset part of the deal? Certainly not the U.S. government, which since 1978 has specifically said it will *not* be involved in these offset arrangements. It is up to the industries to investigate the possibilities for offsets and cooperative industrialization by sending its personnel to the countries of perspective customers. As for the regime rule that says arms suppliers must "export or expire," industry does not need to be prompted regarding the criticality of this systemic rule. Speaking in the aftermath of a recent deal in which South Korea's agreement to buy $600 million worth of P-3 maritime patrol planes kept the assembly line alive, Lockheed president Daniel Tellep called it a "big, big order." Lockheed's future depends very much on being awarded the contract for the new maritime patrol aircraft for the United States. Tellep acknowledged that this contract may not go to Lockheed, but he asserted that the South Korean order "puts a whole new element in the equation."[40]

The general perception among the U.S. defense industrial community is that only those companies who can respond effectively to international trends will survive in the 1990s. This is best seen in analyzing how U.S. defense industries will fare as the European Community moves toward a single integrated market after 1992. My interviews with European and U.S. officials in April 1990 revealed that those firms given the best chance for future orders with European customers are those who have made the effort to establish a presence in Europe, either through mergers and acquisitions or through subsidiaries. Although there are some national-level rules that must be adhered to, none of these efforts are a function of strictly U.S. government actions. In the main, they represent commercial firms responding to international realities.

## *Arms Transfers and Influence: Systemic Pressures Shape the Evolution of an Instrument of Foreign Policy*

The issue of arms transfers and influence provides a final example of how changes in international arms trade systems can explain behavior of actors within the system. As seen in the various regime rules put forward in the three postwar systems, the use of arms transfers as an instrument of foreign policy (particularly by the superpowers) dominated most of the period. But in the 1980–92 system, the utility of this instrument declines. Why? Looking only at the United States, one is hard-pressed to find an explanation at the national level. All of the actors in the bureaucratic politics of arms transfer policy-making, especially the regional and country desk officers in the State Department, consistently press for the use of this instrument. For example, in the spring of 1991, at the peak of the international consensus on the negative consequences of the arming of Iraq, key U.S. Defense Department official Paul Wolfowitz cautioned that "we don't want to construct a regime in which our friends are the principle victims."[41] Some requests by countries for arms may be resisted by the

Defense Department because of the sensitive technology involved. The Treasury Department may object because of negative economic effects on the recipient. Those charged with arms control may object based on the contribution of a transfer to the proliferation of military capability not in the interest of the United States. However, the State Department is still charged with maintaining good relations with all countries, and all things being equal, will generally promote transfers. This is particularly true if the transfer is perceived to increase U.S. influence in the recipient country. Any content analysis of congressional testimony will reveal this orientation, all the more important considering that in the United States, arms transfer policy is managed and led by the State Department, not the Defense Department.

Assuming that the overall goal of U.S. foreign policy continues to be maximizing U.S. influence in the world, why has there been a decline in the use of the arms transfer instrument for this purpose?[42] Those charged with observing the bureaucratic politics of foreign policy-making will not detect a shift in the "organizational essence" of the major actors. Rather, they will find these domestic actors responding to shifts at the international level. As only one example involving arms transfers, the stalemate in Afghanistan is instructive:

> In Afghanistan, for example, the last proxy war of the East-West confrontation revealed how the superpowers' sway has diminished. First, Muslim guerrillas equipped with American weapons humiliated the mighty Soviet army and forced it to withdraw; then they turned on their American benefactors and rejected Washington's advice on a political settlement. Now Afghanistan, one of the poorest countries on Earth, is under the control of neither superpower. "We can't deliver our Afghans," a harried U.S. diplomat said, "and they can't deliver theirs." . . . "The ability of outsiders to influence events is generally declining," said Richard Haass, a former Harvard professor now on the National Security Council staff. "There are simply too many sources of wealth, technology and arms for either the United States or the Soviet Union to be in a position to dictate local decisions. . . . Denial of military or local support is thus a less credible sanction than it was. So is the threat to intervene."[43]

## Continuity, Change, and System Transformation

The above examples have demonstrated that the structure and rules of different temporal systems can be used to explain the behavior of the elements within the system. These elements include national-level suppliers and recipients, as well as subnational actors. Shifts in the basic relationships among actors can also be fruitfully explained at the systemic level, as evidenced by this brief look at how the relationship between arms transfers and influence has changed as different postwar arms trade systems evolved. What remains in this analysis is to address the question of continuity and change. Having delineated these four international arms trade systems and made a case for their utility, what can be said about commonalities? What are the larger, more general conclusions one can reach about arms trade systems?

## The Verification of System Transformation: Sorting Out Continuity and Change

How can one know that a system has been transformed? This is critical if all of this attention to systems is to pay off in forecasting future shifts and transformations. At this point in the book, it is not appropriate to detail the larger debate on this question taking place among international relations theorists. But we can fruitfully explore some of the key points in the debate and apply them to the specific issue area of arms transfers.

Genco explores the issue of systemic continuity and change using the issue area of West European integration. The question for this issue area is not unlike that of arms transfers: how does one know when the movement toward integration has been transformed and is operating under a different set of rules? He comes up with six basic characteristics that distinguish systems transformation from other change processes that are more incremental and focus on the decisions made at the national level. Several are relevant to the study of arms transfer systems, particularly the idea of step-level versus incremental change. "The essential difference is that incremental growth involves changes in amounts and dimensions that are already established; the changes are quantitative, not qualitative. Incremental change can be predicted by *projecting* well-established trends, whereas this is often not possible with step-functional change, for it may involve large and unexpected variations and the introduction of wholly new variables."[44] The three transformations in international arms transfer systems outlined in this book have both quantitative and qualitative step-level changes. The interwar system was changed abruptly by World War II. The bipolar 1945–65 system was transformed by step-level changes in the number of suppliers and recipients, the magnitude of the trade, the switch from giving away surplus material to selling newly produced weapons systems, and the sudden availability of massive quantities of petrodollars in the developing world. The transformation to the 1980–92 system was less dramatic, but the combined effect of the debt crisis in the Third World and the gradual thawing of the cold war starting in the mid-1980s played a major role in the transformation to a system with a clear set of regime rules. The transformation of the 1980–92 system was very clear, more clear than most analysts would have predicted, even as late as the end of 1990. The changes were step-level and quantitative as well as qualitative in nature. A major supplier, the Soviet Union, was literally eliminated from the system, replaced more than likely by China, a second-tier supplier just five years earlier. The primary mode of transfer quickly turned from a trade dominated by political considerations to one more concerned with economic benefits to ailing defense firms. Quantitatively, although data is still preliminary, aggregate global arms exports for 1991 may be down as much as twenty five percent from 1990 levels.[45]

Genco also cites other characteristics of systems transformation: it is a systemwide rather than a sectoral process of change, and external factors play a greater role in systems transformation than they do in other change processes. The external factors used in Genco's account of West European integration were the rejection of the first British bid to join the EEC and the failure to form the European Defense Community (EDC). The former was greatly influenced by French uneasiness with the larger and special relationship between the United Kingdom and the United States, and the demise of the EDC was a function of the relative thaw in the cold war that followed the death of Stalin. Both of these were systemic factors, and as has been shown, similar external factors were the keys to the transformation of the postwar arms transfer systems.

Rosenau points out that there are two polar extremes in dealing with continuity and change. One extreme

> stresses the openness of social systems to redirection and transformation. It tends to equate crises with breaking points, to view tensions in world affairs as perturbations that herald important changes. It emphasizes the susceptibility of systems to shocks of violence and upheaval. At the other pole is a tendency to perceive history as a seamless web in which there are no breaking points. This perspective stresses the power of habit, the pervasiveness of cultural norms, and the constraints of prior experience as predisposing systems toward continuity.[46]

Rosenau goes on to propose that those seeking to explain change, or system transformations, need to search between these extremes. Several of the assumptions underlying this between-the-extremes approach are very relevant to the systems approach taken in this study of international arms transfers. One assumption is that "the interpretation of continuity and change depends on the systemic and time perspectives from which they are assessed. . . . It is not history that dictates whether change has occurred, but rather the interests of observers, the scales of time and space in which they seek to trace changes in the past and to evaluate those that may lie in the future."[47] The implications of this assumption for a systemic-level analysis (such as is contained in this book) is that system transformation is defined by its utility to the observer—which is the case with any scientific enterprise that employs concepts, whatever the level of analysis. The previous section of this chapter was designed to make exactly this point, that the systems put forth represent the reality of arms transfers as perceived by the major actors in the system.

## Continuities in Historical International Arms Trade Systems

This first assumption led Rosenau to state a second: "The longer the time span and the more encompassing the system, the greater the probability that the statics of continuity will prevail over the dynamics of change." Several recent studies of the international arms trade at the systems level are examples of this assumption at work. Kolodziej and Pearson have put forward a framework built around two competing systemic needs—order and welfare. "One place to begin building such a body of testable propositions about nation-state and market behavior is with the making and marketing of arms.... Arms production and sales combine both the incentives for behavior derived from command and coercion with those arising from consent and exchange for mutual benefit."[48] This study proceeds to provide preliminary evidence as to how these two imperatives interact by outlining the behavior of key states in the system, not unlike the exercise conducted in the previous section of this chapter. Among other things, they confirm that these forces operate on the second-tier suppliers (Brazil, Israel, and others). The imperative to have order forced Israel to create its own defense industry, whereas the welfare imperative (that Israel is subject to the basic laws of the market irrespective of its security demands) has resulted in a significant decline in Israeli defense exports and overall indigenous production.

Kolodziej and Pearson's conclusions are several. First, "neither the paradigm and postulates associated with strategic-diplomatic nation-state behavior nor with the developmental or dependency theories of the market fully explain the international arms economy."[49] Secondly, "as the costs of supporting MISTS (military-industrial-scientific-technological systems) and national arms production have risen, most arms producers have increasingly emphasized the commercialization of their arms, services and know-how to sustain their national efforts."[50] Having put forward the general proposition that these two imperatives matter, they conclude that "if progress is to be made either in explaining state behavior or in fashioning policy instruments to regulate the struggle for power and welfare, realist assumptions that differentiate the theory of the state from the market as well as order from welfare will have to be revised to integrate these domains within a common conceptual framework."[51]

The Kolodziej-Pearson study points out several things related to the systems approach to explaining international arms transfer behavior. First, it did not involve the delineation of time periods and, in essence, treated the entire postwar system as one system. This tends to place it toward the extreme pole defined by Rosenau as being dominated by historical evolution. Although the basic truth of these two imperatives impinging on the behavior of actors in the system is revealed, it is difficult to go beyond that basic point. Secondly, as one attempts to elevate theory and explanation to a higher plane, the ability to explain and forecast specific behavior declines. Kolodziej and Pearson were obligated to expand the time period of their observations in order to allow the evolution of history to bring forth enough variations in national behavior to make their basic point. However, in doing so they lose the specificity that comes with observing phenomena in shorter, more discrete time periods. Both intellectual enterprises are essential as we seek to define arms transfer behavior accurately; they are, however, quite different in their application to the policy world of national-level decision makers.

The recent work of Keith Krause also makes clear the implications of expanding the temporal dimensions of arms transfer systems. His study starts with the emergence of the modern state system in the sixteenth century and traces developments up to the current set of upheavals described in this book. The study "argues that a prerequisite to the understanding of micro-level (interstate) developments in arms transfer relationships is a 'systemic' perspective of change and transformation of the international arms transfer system." He puts forward four propositions:

1. The international arms transfer system is subject to a series of cyclical "technological revolutions" that are catalyzed by the imperfect diffusion of a new technology throughout the system.
2. As the system evolves, it manifests a consistent three-tiered structure that is obscured during periods of rapid technological change (after a technological revolution).
3. The current (post-1945) system is located in one of these periods of rapid change, but is evolving towards a more "normal" structure.
4. The future evolution of the current system will follow previous patterns of imperfect diffusion for both the technologies and the techniques for production of arms.[52]

Treating the entire period of history as one system, Krause develops five main systemic characteristics. The first four are the emergence of leading centers of production and innovation, a growing political salience and impact of the arms trade at both the level of state policies and the systemic level, the diffusion of military techniques that grow out of the political salience of the arms trade, and the imperfection of the diffusion process (that is, the attempts at technology transfers that fail to bear fruit in their new soil). His fifth, and perhaps most important finding, was that the arms transfer system developed a coherent structure, based on a division between producers and consumers and a three-tiered division within the ranks of producers. The first-tier producers innovate, the second tier imports the capacity to produce, and the third tier copies technologies but does not capture the underlying process. The consumers in his system obtain only the material transfers of the tools of war.[53]

Krause's major finding is that the 1945–65 period stands out as an historical aberration, since several expected participants were eliminated by World War II (including Germany and France), the USSR remained out of the system until 1955, and the third tier of states did not develop military industries "due to the legacy of colonialism and the inward focus of post-independence politics."[54] Rosenau's point about the temporal dimension influencing the findings of systemic-level research are made perfectly clear in Krause's key critique of contemporary studies of the arms trade:

These aberrations resulted in an artificially distorted oligopolistic system based on American dominance (and rising Soviet participation) as the model on which most analyses of current changes (late 1980s) are based. A longer historical perspective would have immediately suggested that the post-1945 situation was temporary and likely to change as something more closely resembling the traditional equilibrium re-emerged. This has indeed occurred over the past two decades, and it reflected not idiosyncratic (and "threatening") decisions on the part of the Soviet Union, France, Germany, Brazil, or India but the expected response of nation-states to the incentives to produce arms which are built into the international system.[55]

Krause's findings based on the longer historical approach validate the utility of the systemic approach and also clearly point out the implications of choosing specific temporal boundaries. On the one hand, an overemphasis on the permanence of the cold war, which failed to recognize that the dominance of the United States and the USSR was temporary, to explain arms transfer behavior may have held sway long past its time because of the choice of temporal boundaries. On the other hand, Krause's findings are at such a level that it will be difficult to use them to forecast systemic shifts in the timely manner demanded by national-level decision makers. And his approach still requires delineating when and under what conditions the system will return to "normal." However, the longer historical view serves as an invaluable baseline against which to assess the forecasts of experts. Indeed, Krause cites many examples of the assessments of others that are simply wrong, and the contribution of such critiques should not be underestimated. His three-tiered structure conforms to Neuman's major research into the stratification of production[56] and, despite being based on the longer period, is extremely useful as we observe the behavior of second- and third-tier states in the current upheaval that is producing declining national defense budgets. Krause supports one of the major findings of this book, that arms transfer relationships as a tool of foreign policy are in decline. Regarding control of the arms trade, he predicts that progress is unlikely until the system returns to "normal."

Both the Krause and Kolodziej-Pearson studies hold that there are major continuities in the system. However, this is not to say that continuities cannot be discovered when one analyzes the arms trade using the shorter time periods found in this book. For example, once the nation-states of the interwar system agreed to national control of arms exports, this systemic characteristic held despite other major changes in the system. This has resulted in defense industries behaving in an amoral fashion, letting national governments decide what is right and wrong. As the president of an American electronics firm recently put it when defending sales of computers with military potential to Iraq, "Every once in a while you kind of wonder when you sell something to a certain country. But it's not up to us to make foreign policy."[57] As an example of another continuity, the motives for acquiring arms do not seem to have changed much as the temporal systems change, particularly for those states engaged in conflict. This demand factor will mean that an international arms trade system of some kind will always be operating. Actors in the system, be they industries trying to boost the bottom line or nation-states trying to exert influence, will always seek to respond to this systemic characteristic. Other continuities can be revealed by comparing the summaries of each system, particularly the regime rules.

## Sources of System Continuity and Transformation and the Post-1992 Arms Trade System

Another advantage of the more frequent and shorter time-period approach used in this book is in its ability to forecast future systems. Krause may be correct in his finding that the current system is evolving to his baseline. But his effort to provide a more general and basic explanation of the system, an effort that has been most successful, has resulted in a reduced set of systemic variables, particularly when compared to those used in this book. It comes down to asking what one watches and monitors while the transformation is taking place. The system characteristics and regime rules used in this book provide a rich menu of such variables to monitor, and give the observer more clues regarding systemic transformation. Also, to the extent that the regime rules for each system are derived from the other categories of variables, future regime rules become easier to predict.

It is clear from the analysis of the characteristics of the 1980–92 system that many things will be different in the post-1992 arms trade system. To focus on those changes, it is useful at this juncture to ask how the previous systems maintained their stability. What was it that enforced the "rules" of the system? As we have seen in the three systemic transformations described in this book, continuities exist from system to system. Which systemic characteristics are likely to continue from the current system into the post-1992 international arms trade system, and which ones will be different? How will system and regime maintenance differ in the post-1992 period? To answer these questions, sources of continuity and change for each of the systemic categories used in this book will be addressed.

### Boundaries and Environment

It is fairly certain that the bipolar political system that shaped the three postwar arms trade systems will not return. The security landscape of Europe now finds Poland, Hungary, and Czechoslovakia working hard to join NATO. Within the former Soviet Union, economic, political, and even military chaos insures that most of their energies will be directed inward for the foreseeable future. While some arms exports are bound to occur, given the lingering dominance of military industries, it is highly unlikely that arms exports will return soon to anything approaching levels found in the 1980-92 system, either by Russia, Belarus, or the Ukraine, where most of the armaments manufacturing plants are located. Indeed the elimination of politically motivated exports from the Soviet Union has eliminated much of the reason for the United States to respond in kind. As a result, in the near term, the arms trade will be unhooked from the larger international political system, a system which itself is still evolving.

The trend toward regional conflicts generating a demand for conventional armaments that developed in the 1980-92 system will continue, perhaps even increasing with the rebirth of nationalism and ethnic strife unleashed at the end of the cold war. Such a system will promote the attractiveness of suppliers such as China and perhaps North Korea, since their state-controlled military industries produce the type of combat-tested, mid-level technology, and low-cost weapons systems needed to fight the type of nationalistic and ethnic conflicts that appear to be ascendant in the system. But without the East-West rivalry in military technology, the drive toward more sophisticated weapons systems will be dampened, and recipients of arms transfers will concentrate more on absorbing current technologies and inventories.

Economically, the massive nonmilitary requirements that surfaced in the 1980-92 system will continue to dominate the next system. Money for big-ticket military items such as the latest tank, fighter aircraft, and missile will be hard to come by for both developing and industrialized states. In this environment, creative financing will continue to be required. Even a return to high oil prices is not likely to result in a buying spree that could allow recipients to buy for prestige reasons, or defense industries in industrialized countries to use increased arms exports to ward off a reduction in size. The new system will feature a much smaller global military-industrial complex as a result of significantly reduced defense budgets in industrialized states and the presence of massive quantities of usable military surplus equipment being available for trade.

One international geopolitical factor that began to emerge in the early 1990s may become crucial to the post-1992 arms trade system. This is the growing trend toward regionalization of the international economic system. The move toward a single market in Europe will, at least in the early part of the new system, result in a "Europe first" approach to trade. This can be seen in the current impasse in the GATT negotiations, a direct result of Western Europe refusing to give up subsidizing its agricultural production. To the extent that this trend continues, it will also force defense industries to consolidate along European lines. The United States, Canada, and Mexico are moving toward a North American free-trade regime. In Asia, the recent Association of Southeast Asian Nations (ASEAN) meeting called for further economic integration. Given the increasing commercialization of the arms trade, it is likely that arms production will also become regionalized.

## Characteristics of Units

In light of the negative international reaction to the revelations of illegal arms trade in the 1980-92 system and the role of private firms in the arming of Iraq, it seems likely that the norm that developed in the 1930s regarding national governments' control over the export of armaments will continue after 1992. Nevertheless, the trends emerging in this current system toward commercialization, the internationalization of defense production, illegal arms trade, and trade in dual-use and upgrade packages will continue to put pressure on these governments to control the negative consequences of the arms trade. The post-1992 system will feature industrialized states acquiring arms from their regional partners to meet the objectives of defending their sovereignty and meeting their regional (and perhaps international) security responsibilities. As for developing states, the lack of money for acquiring new systems will mean a continuation of the trend toward upgrades and an increase in the acquisition of used-but-modern weapons systems. Multinational defense industries will remain in the system, but the national control norm will prevent a total return to the interwar system in which these firms were very influential in determining conflict outcomes.

## Structure and Stratification

The potential for systemic change is the greatest in this area, although some things will look the same. For example, the system will continue to be a buyers' market, and most of the trade will be to developing states engaged in regional conflicts. It also appears that the Middle East will continue to dominate the trade, given the poor prognosis for conflict resolution and the likelihood that the industrialized world will remain dependent on oil. Second-tier suppliers will continue to exist, particularly given their niche as suppliers of mid-level technology weapons systems for use in regional conflicts.

A major factor inducing change will be the reduction in the number of suppliers of advanced weapons systems. In Western Europe, the cross-national consolidation and integration of defense industries that commenced during the 1980-92 period is proceeding at a rapid pace. There are significant efforts under way to integrate West European defense industries, leading some to predict a politically united and economically integrated Europe that will behave as one major arms supplier in the system.[58] It is unlikely that systems such as main battle tanks and advanced fighter aircraft will be produced by individual European nation-states. It is more likely that a pan-European tank and fighter aircraft will be produced, significantly reducing the number of suppliers in the system. The decline of the second-tier suppliers that became a reality in the 1980-92 system will not change, despite the fact that observers of the current arms trade system and forces within these second-tier suppliers continue to project that as the industrialized arms suppliers cut back, the second tier will become more competitive.[59] The actual role of these suppliers will depend greatly on the development of arms trade control regimes in the post-1992 system.

As this book is published, the Soviet Union will have been absent as a major arms supplier for about one year. The raw empirical data has begun to verify this decline, but there is a great deal of evidence that even had the Soviet Union survived a major systemic transformation would have occurred. Its once-feared military-technology prowess was already very much in decline when it disintegrated. As only one example, efforts by the USSR to develop cooperative space programs with the United States, once seen as a major route to improving U.S.-Soviet relations, are being resisted by a United States that is questioning the reliability and technological capabilities of the USSR.[60] Additionally, the performance of Soviet equipment in the recent Gulf War has raised questions in the mind of many potential customers.[61] The Soviets put forward a steady stream of arms trade control proposals,[62] and were cooperating with the Missile Technology Control Regime.

But what of the successor republics of Russia, Belarus, and the Ukraine, all of which have major conventional arms production capabilities? How will they participate in the post-1992 arms trade system? While it is still very early for definitive answers, some trends are emerging. First, while little hard evidence ever emerged during the cold war regarding the arms export policymaking process, it was always assumed that the decisions to export arms were very political and very centralized. This assumption has been confirmed in the wake of the collapse of the Soviet Union, which provided that central control. This central control performed three functions that are now missing. First, it coordinated arms production, across defense firms and in many cases across republics. In the absence of such control, it is now common to observe Russian defense plant managers touring Western Europe and the United States, looking for defense contracts.[63] A second function of this central control was the promotion, arrangement, and execution of arms deals. Without such control, we have seen the result—a haphazard, almost desperate, sales effort. This includes everything from appearing at exhibitions to a host of amateur arms brokers trying to sell almost anything to anyone. In the former system there was no shortage of international arms sales efforts, but it was done through a vast political and tightly controlled network.

The current frenzy also includes inviting anyone interested in either buying or coproducing military equipment to Russia to tour various facilities. This created an interesting problem for the United States in the fall of 1991. A major U.S. goal is to insure that the republics of the former Soviet Union reduce the size of their military industries through conversion, a task that has been going on for several years.[64] In the fall of 1991 a team of U.S. industrialists accompanied Deputy Secretary of Defense Donald Atwood to Russia to assess the conversion effort. They were greeted with myriad requests not for help with conversion but rather with requests to coproduce military equipment with the United States. As a result of some apparent enthusiasm at the Defense Department for this idea,

Atwood slapped a moratorium on any contact by U.S. Defense Department officials with Russian military industry officials, lest the United States unwittingly derail the Russian conversion effort.[65]

A third function of the centralized Soviet arms export system was the control of exports. With its dissolution went all of the laws governing conventional arms exports, with little evidence of anything similar taking their place in the successor republics. As a result, it is difficult to know what exports have taken place in the past year. Additionally, there is the often stated fear that exports will naturally increase in such a lawless environment. Regarding nuclear weapons, these fears are well founded.[66] But up to this point it can be said with some confidence that conventional weapons and the scientists and technicians who make them have not had much success in exporting their wares and talents. The difference lies in systemic demand. Any number of budding would-be nuclear powers are anxious to pay top dollar for the nuclear technology and weapons that have suddenly become more available in the former Soviet Union. But the demand is very different for conventional weapons, i.e., a demand that has been steadily declining since 1989. It is true that the incentive for arms firms everywhere in the system to export or expire has never been greater. There are simply very few customers willing to pay the money that these firms need to stay in business. What then are these recipients using for equipment? Sometimes they make do with what they have, other times they upgrade, or, increasingly, they buy almost-new equipment made surplus by arms control agreements or national-level defense budget cuts. In short, the theme of this book, that systemic characteristics determine national outputs, appears to be very much at work in explaining the lack of significant arms exports from Russia and the Ukraine.

In regard to the proliferation of military power via the arms trade, the experience of those states that acquired and utilized advanced weapons systems in the 1980–92 system may introduce a new reality to the post-1992 system. It may be that these states, having seen the effect of using such power to threaten the industrialized states, may be willing to show some restraint on the demand side of the equation, particular if some rewards are attached. This appears to have been the case when the United Nations adopted the Transparency in Armaments resolution 150-0 in December 1991. Traditionally, developing countries showed the least amount of support for arms trade control. Why the sudden reversal on the arms trade register? One reason might be that they saw that arms buildups have their limits and can cause an international consensus for action, e.g., the embargo against Iraq. However, it is more likely that they went along on a trial basis to test the major suppliers' willingness to declare exports. This could be of significant benefit to a country like Egypt, which might find it useful to know what has been imported by Israel. Additionally it should be noted that the price for getting the votes of some key developing countries was the convening of a panel in 1994 to assess the first two years of the register and consider expanding the register to include high technology and weapons of mass destruction.[67]

Developing country restraint in acquiring ballistic missiles appears to be more likely than other conventional weapons systems. If it was the threat of using ballistic missiles with warheads of mass destruction by Iraq that sparked the military response of the coalition forces, as argued in this book, recipient states may be more willing to cooperate in controlling the trade in such systems.

Additionally, the dampening of the military-technology race between the United States and the USSR may lead to a lowering of demands for the very latest technology and satisfaction with advanced but usable weapons systems. This seems to have been the case with the recent fighter aircraft acquisition by South Korea. Initially, the F/A-18 was selected in 1989. However, because of a price increase by the manufacturer (McDonnell-Douglas) and some foot-dragging by the U.S. government regarding the technology to be transferred, the South Koreans changed their minds and will now be acquiring and coproducing the F-16. This would seem to be a case where the recipient decided to forgo the more expensive and technologically advanced aircraft for one that was very modern and adequate to meet foreseeable threats (in this case, on the Korean peninsula).[68] One of the major lessons learned from the 1980–92 period, especially in the Gulf War, is that the acquisition of the most sophisticated weapons systems by developing states does not always translate into national security.

## Modes of Interaction

Many of the trends in modes of payment and production that emerged in the 1980–92 system will continue past 1992. For example, the trade will be very commercial in nature. But there is some evidence that the use of offsets will not be as prevalent in the post-1992 system. First, there is a backlash occurring in those industrialized states who have engaged in this mode of interaction in the 1980s. With defense industry employment in sharp decline, exporting jobs via offsets is becoming a political liability. The 1990 Office of Management and Budget report on offsets deals with this question in great detail. This same report also makes the point that as the arms trade system turns more toward multinational production, offsets may decline in favor of sole-source procurements from newly developed regional defense industries.[69]

International production will play a much larger role in the next system. As indicated in the above discussion of structure and stratification, pan-European production will be more important than the nationally produced weapons that have dominated previous systems. This internationalization may go beyond regions. In a recent interview, Gordon Adams of the Defense Budget Project in Washington gave the following mythical—yet quite probable—example: "The

General Dynamics tank of the future may well be designed jointly with Vickers in Britain, prototyped in Lima, Ohio, assembled in Delhi, Tel Aviv and Cairo, with a digital firing system manufactured in Tokyo. Is that an American tank? You tell me."[70]

*Regime Norms and Rules*

**Conflict.** The regime rules that described the relationship between arms transfers and various stages of armed conflict will continue to hold in the post-1992 period. It will be very difficult to control the outbreak, pace, or termination of conflict in an arms trade system even more open to the acquisition of military capability that can be used in regional conflicts between developing states. In regard to developing states acquiring enough military capability to threaten the major powers, much will depend on the arms trade control that emerges out of the current flurry of activity on that front. If the major supplier states in the system forgo the opportunity to control negative consequences such as seen in the case of Iraq, it is highly likely that developing states will continue to present problems for these suppliers in the post-1992 system.

**Foreign Policy and Diplomacy.** The essence of the regime rules on this dimension for the 1980–92 system is that gaining influence through the use of arms transfers is extremely difficult. It is unlikely that enough change will occur to alter these regime rules in the post-1992 system, especially given the elimination of the USSR as an arms supplier. Arms that are transferred will remain at a low level of political visibility, be they upgrades, dual-use or used equipment, or systems produced by apolitical international firms unresponsive to national and international political pressures. It will be difficult for the suppliers in the system to reach any meaningful consensus to effect a change in these regime rules. In essence, the post-1992 system may resemble the interwar years, when arms transfers had little to do with foreign policy.

**Economic Effects.** The export-or-expire rule will decline in importance as the shakeout of defense industries is completed sometime before the end of the decade. For example, in Europe, the current system dictates that arms producers in France (including government-owned firms) continue to think first of exports, since an already-small domestic market will be at least 25 percent smaller by 1995. This may not be the case as Europe integrates and rationalizes its defense industries. One of the major goals of the Single European Act is to expand the size of the regional European market so that European firms will not have to rely as much on exports outside Europe. This could very well happen with defense production, although many taboos, national cultural preferences, and laws will have to be changed for this to happen. In the United States, the rationalization of defense industries is well under way, as the end of the cold war has taken away the last excuse many industries had to hang on for the big upswing in orders. In short, the domestic markets of most suppliers—perhaps redefined to include regions—will mean that there will be less economic incentive to export to the international system as a whole. This will also mean a decline in the use of creative financing, with more emphasis on pay-as-you-go policies. Those states with critical national security problems will find many sources of inexpensive upgrade packages and usable surplus equipment. Without the cold war driving technology to its limits, particularly for conventional weapons, the cost of defense (or aggression, for that matter) will be much lower. The merchants-of-death syndrome will continue to disappear from the international scene, unless it is integrated into one of the arms trade control regimes that may surface.

One new regime rule may emerge. For the past two arms trade systems, it was assumed by developing states that the model for economic and industrial development included acquiring a military-industrial capability as a first step. Despite economists who tried to prove otherwise, this was a popular model and was responsible for much of the offset and technology-transfer activity in the current system. However, a sober assessment by many of these countries may demonstrate to them that this was a false approach, that in fact the conversion from a fledgling defense industry based on coassembly and coproduction to purely civilian industry is costly and technologically complex. The new model, which is being increasingly supported by large national and international financial donors, holds that it is more efficient to develop civilian industry directly. This will directly affect arms trade patterns, with more states beginning to buy off-the-shelf items rather than coproducing them at much greater expense. Jobs, including those in defense industries, will still be important and will be the major factor in shaping offset packages; it is the technology-transfer dimension that may decline in importance. Leading the way in this new way of thinking about industrial development will be Europe, already far along in emphasizing civilian over military industrialization.

**System Maintenance and the Control of the Arms Trade.** One of the major effects of the arming of Iraq and the subsequent Gulf War is the sudden increase in the attention paid to system maintenance and stability through the instruments of multilateral arms control. As pointed out when describing the three postwar arms trade systems, these systems were maintained in essence by the larger political system, in which the superpowers exported arms in a manner that theoretically did not create destabilizing regional or global conflicts. A stable system was one in which the negative consequences of the arms trade to the major powers, especially the superpowers, were acceptable. Both superpowers practiced what amounted to unilateral arms control and tacit cooperation. The 1980–92 system, which featured the Afghanistan, Iran-Iraq, Iraq-Kuwait, and Persian Gulf wars, demonstrated that such maintenance techniques were becoming outdated. As of 1992, it cannot be said that any other system-maintenance procedures and processes have emerged to take the place of this superpower control. However, what has emerged is an unprecedented number of approaches and concrete proposals to control what is now perceived to be the dangerous proliferation of military capability that threatens the stability of the system.

The first of these trends is that of *transparency*, the opening up of information on the arms trade so as to allow the affected states to dampen and eliminate the negative consequences that ensue. Some of this transparency was unintended, such as the unwanted publicity that Germany received as a result of transfers to Libya and Iraq. Lists of firms and of the items exported that led to undesirable military capability in these two states provided the most thorough evidence made public as to how a developing state can acquire the ability to produce ballistic missiles with warheads of mass destruction.[71] Even so, transparency also began to surface as a purposeful effort. In the spring of 1989, the UN sponsored a conference on the subject in Italy and published the papers in 1990.[72] In the spring of 1991, in the aftermath of the Gulf War, country after country began to publish details of its arms exports and to put forth proposals for transparency and the idea of an international arms trade register. The French, Germans, Bulgarians, Czechs, and the Soviets all published heretofore unreleasable data on arms exports.[73] As the July 1991 G7 summit meeting approached, the leaders of Japan and the United Kingdom put forth formal proposals for an international arms trade register, an idea that was ratified at the actual summit.[74] These proposals were merged and became the proposal adopted as General Assembly Resolution 46/36L in December 1991. As previously mentioned, the arms trade register exists and a panel of experts is finalizing the procedures that will govern its operation. The first voluntary submission of national data is scheduled for April 1993.

But proposals for arms control were not limited to the concept of transparency and an arms trade register. There was an unprecedented outpouring of arms trade control proposals from defense trade publications,[75] the CEO of Daimler-Benz,[76] supplier governments,[77] recipient governments,[78] and European organizations.[79] Contained in these proposals were policies to tighten up export procedures and to begin developing more international controls. On May 29, 1991, President Bush announced the "Middle East Arms Control Initiative," calling for the five largest arms-supplying nations (the United States, the USSR, France, the United Kingdom, and China) to meet in Paris to "establish guidelines for restraints on destabilizing transfers of conventional arms, as well as weapons of mass destruction and associated technology." The proposal also called for expanding the talks to other suppliers and for permitting states in the region to "acquire the conventional capabilities they legitimately need to deter and defend against military agression." To implement the regime, suppliers would commit to "observe a general code of responsible arms transfers, avoid destabilizing transfers, and establish effective domestic export controls on the end-use of arms or other items to be transferred." The proposal also called for a consultative mechanism. Further, it was recommended that a freeze be put on surface-to-surface missiles in the region, with a goal of their eventual elimination.[80] Complementing and sometimes challenging the president's effort are a series of proposals from the U.S. Congress that range from an outright ban on all arms sales to the Middle East to support for the arms register concept.[81]

If one were to list all of the relevant op-ed pieces written in the spring of 1991, it would only confirm the salience of the issue of controlling the arms trade. What is to be made of this phenomenon? First, the attention in the United States looks very similar to that of the mid-1970s, which led to stiffer arms export-control legislation and the Carter arms transfer restraint policy but not much of an effect on U.S. exports or the international arms trade. This time around, however, there is not only international attention to the issue but also a consensus in the various proposals that certain types and levels of arms transfers can lead to negative consequences for the industrialized world. As of March 1992, the five permanent members of the UN Security Council had met three times on the issue. If this and other negotiations and discussions lead to a new type of multinational control regime, perhaps some type of Middle East accord or an enhanced MTCR, it may provide the system maintenance for the post-1992 period. However, the process may get stalled. Without a bipolar structure to maintain stability, arms exports would begin to resemble tractor exports in terms of their availability and accountability. At that point, the world may have to wait for another demonstration of the negative consequences of the arms trade to reinvigorate the arms trade control process. Alternatively, the economic situation and the demand for solutions to nonmilitary problems may provide another type of control, preventing the acquisition of dangerous capabilities by restricted access to the funding needed for advanced systems.

## The Next Steps in Explaining the Arms Trade

The major theme of this book has been that the first and most important source for understanding the international arms trade is evidence at the systemic level. Four different systems were described, along with the basic reasons for their transformations. In an attempt to emphasize this level of analysis as the first step in understanding the arms trade, claims for the utility of the approach may have been exaggerated. As chapters 1 and 2 indicate, a significant amount of knowledge of the arms trade exists at the national and subnational levels. And many questions cannot be answered at the systemic level. For example, *within* a particular historical system, what accounts for different behaviors on the part of suppliers and recipients? Questions like this call for a rigorous and comparative assessment designed to generate the same type of generalizations found in this book at the systemic level.

There is much work that needs to be done on these other levels. Although many case studies of suppliers and recipients exist, few genuinely comparative studies that lead to policy-relevant theory have been undertaken.[82] Some of the best work has been done by SIPRI; for example, their comprehensive 1987 study, *Arms Transfers to the Third World: 1971–85,* has a chapter dedicated to suppliers that asks the same questions of each case regarding the flow, institutional framework, arms production, and political issues. But no attempt is made to aggregate the findings to generate general behavioral rules. The findings from both the Krause and Kolodziej-Pearson studies could be easily adapted to a focused comparison of suppliers. Kolodziej's study of the making and marketing of French armaments remains the standard for case studies that generate propositions that could be used in a comparative study. It is particularly good at showing how French national forces interact and respond to the changing international system. Once comparative studies of actors have been completed, another step would be to use data from the national level to analyze relationships across systems. In this book, for example, it was shown how shifts in the relationship between arms transfers and influence parallel and are explained by shifts in the larger international arms trade system. Neuman's study of the effects of arms transfers on conflict is another example of how this type of study can generate policy-relevant theory.

This book has shown that explaining the international arms trade can be accomplished at the international systems level. But as with any other international issue area, fully explicating this critical instrument of foreign policy can only be done at multiple levels of analysis. It is hoped that the systems-level approach of this book will serve its purpose of providing the starting point and baseline to which other levels of evidence can be added in order to bring us closer to a true understanding of the patterns, purposes, and effects of the international arms trade.

# Notes

## Chapter 1: An Introduction to the Study of International Arms Transfers as a Concept and Issue Area

1. Ernst-Otto Cziempel and James N. Rosenau, *Global Changes and Theoretical Challenges* (Lexington, Mass.: Lexington Books, 1989).
2. Amelia Leiss et al., *Arms Transfers to Less Developed Countries*, Report No. c/70-1 (Cambridge: Center for International Studies, MIT, 1970), 13.
3. John L. Sutton and Geoffrey Kemp, *Arms to Developing Countries*, Adelphi Papers, no. 28. (London: Institute for Strategic Studies, 1966).
4. John Stanley and Maurice Pearton, *The International Trade in Arms* (New York: Praeger Publishers, 1972), 7.
5. See Aaron Karp, "The Frantic Third World Quest for Ballistic Missiles," *Bulletin of Atomic Scientists* (June 1988): 14–19.
6. There is a rich history of cartoons that effectively capture the history and major controversies of the arms trade from the 1930s on. The author's collection tends to confirm that it is the publicly perceived negative consequences that provide the occasion for the publication of most of these cartoons.
7. Art Buchwald, "Lean Times at the Arms Bazaar," *Monterey Herald*, 27 September 1990; idem, "Armed and Ready, Once Again," ibid., 5 March 1991; idem, "Banner Year for U.S. Weaponry," ibid., 9 July 1991.
8. As only one example, see U.S. Senate, *Foreign Assistance Authorization: Arms Sales Issues, Hearings Before the Committee on Foreign Relations* (Washington: GPO, 1975). One particularly influential report was the report of a staff trip to Iran to investigate the problems associated with U.S. arms sales to Iran; Committee on Foreign Relations, U.S. Senate, *U.S. Military Sales to Iran*, Committee Print (Washington: GPO, 1976).
9. For example, see Uri Ra'anan et al., eds., *Arms Transfers to the Third World: The Military Buildup in Less Industrialized Countries* (Boulder, Colo.: Westview Press, 1978). This book is a compilation of papers given at a conference sponsored by the Fletcher School of Law and Diplomacy, Tufts University, in the spring of 1976.
10. For a brief treatment of this policy, see Andrew J. Pierre, *The Global Politics of Arms Sales* (Princeton: Princeton University Press, 1982), 45–61.

11. The author served in the office of the chief of the Arms Transfer Branch, Arms Control and Disarmament Agency, from the summer of 1978 to the summer of 1979.

12. Jo L. Husbands and Anne Hessing Cahn, "The Conventional Arms Transfer Talks: An Experiment in Mutual Arms Trade Restraint," in *Arms Transfer Limitations and Third World Security,* ed. Thomas Ohlson (New York: Oxford University Press, 1988), 110-25.

13. For contrasting assessments, see Nicole Ball and Milton Leitenberg, "The Foreign Arms Sales of the Carter Administration," *Bulletin of the Atomic Scientists* (February 1979): 367-72; and Bernard A. Schreiver, "Jimmy Carter's Arms Transfer Policy: Why It Won't Work," *AEI Defense Review* 2 (1978): 16-28.

14. For example, see Barry M. Blechman, Janne Nolan, and Alan Platt, "Pushing Arms," *Foreign Policy* 46 (Spring 1982): 138-54.

15. Pierre, op. cit.

16. Christian Catrina, *Arms Transfers and Dependence* (New York: Taylor and Francis, 1988).

17. With the exception of those sales that became known as a result of the Iran-Contra operation, as well as illegal sales. Regarding the latter, see Edward J. Laurance, "The New Gunrunners," *Orbis* 33, no. 2 (Spring 1989): 225-38.

18. Frederick J. Hollinger, "The Missile Technology Control Regime: A Major New Arms Control Achievement," in *World Military Expenditures and Arms Transfers 1987* (Washington: U.S. Arms Control and Disarmament Agency, 1988), 25-27.

19. Robert E. Harkavy, *Arms Transfers and the International System* (Cambridge: Ballinger Publishing Company, 1975), 12.

20. J. David Singer, "The Levels of Analysis Problem in International Relations," in *The International System: Theoretical Essays,* eds. Klaus Knorr and Sidney Verba (Princeton: Princeton University Press, 1961).

21. Amelia Leiss et al., op. cit.

22. Stockholm International Peace Research Institute, *The Arms Trade with the Third World* (New York: Humanities Press, 1971).

23. Ibid., v.

24. For a critical review of this book, see Geoffrey Kemp, "Strategy, Arms and the Third World," *Orbis* (Fall 1972): 809-16.

25. Stanley and Pearton, op. cit.; Sutton and Kemp, op. cit.; Lewis A. Frank, *The Arms Trade in International Relations* (New York: Praeger Publishers, 1969).

26. Ulrich Albrecht, "The Study of International Trade in Arms and Peace Research," *Journal of Peace Research* 9 no. 2 (1972): 165-178; Amelia Leiss, "Comments on 'The Study of International Trade in Arms and Peace Research,'" *ibid.,* 179-82.

27. Harkavy, op. cit.

28. Thomas Kuhn, *The Structure of Scientific Revolutions* (Chicago: University of Chicago Press, 1962).

29. Robert T. Holt and John E. Turner, *The Methodology of Comparative Research* (New York: Free Press, 1970) summarizes Kuhn in a very useful manner. What follows relies heavily on this work.

30. Edward J. Laurance and Ronald G. Sherwin, "Understanding Arms Transfers through Data Analysis," in *Arms Transfers to the Third World: The Military Balance in Less Industrialized Countries,* ed. Uri Ra'anan et al. (Boulder, Colo: Westview Press, 1978), 88.

31. Michael Brzoska, "Arms Transfer Data Sources," *Journal of Conflict Resolution* 26 (March 1982): 78-79.

## Chapter 2: The Conceptualization and Measurement of International Arms Transfers

1. Ian Anthony et al., "The Trade in Major Conventional Weapons," in *SIPRI Yearbook 1991* (Oxford: Oxford University Press, 1991), 202-3.

2. Kenneth R. Timmerman, *The Poison Gas Connection: Western Suppliers of Unconventional Weapons and Technologies to Iraq and Libya* (Los Angeles: Simon Wiesenthal Center, 1991); Gary Milhollin, *Licensing Mass Destruction: U.S. Exports to Iraq, 1985-1990* (Washington: Wisconsin Project on Nuclear Arms Control, 1991); Kenneth R. Timmerman, *The Death Lobby* (Boston: Houghton Mifflin Company, 1991).

3. Michael A. Ottenberg, "Operational Implications of Middle East Ballistic Missile Proliferation," *Defense Analysis*, 7, no. 1 (1991): 3-19.

4. Janne Nolan, *Trappings of Power: Ballistic Missiles in the Third World* (Washington: Brookings Institution, 1991), 68.

5. *The Military Balance* (London: International Institute for Strategic Studies, annual publication).

6. In the course of reviewing arms transfer data in this chapter, all of the key works in this debate are referenced. For the most recent treatment of these issues, see Edward J. Laurance and Joyce A. Mullen, "Analyzing Arms Trade Data," in *Marketing Security Assistance*, ed. David J. Louscher and Michael D. Salomone (Lexington, Mass.: Lexington Books, 1987), 79-98; and David J. Louscher and Michael D. Salomone, "The Imperative for a New Look at Arms Sales," ibid., 13-40.

7. Jonathan Fuerbringer, "Rusty Statistical Compass for U.S. Policy Mappers," *New York Times*, 30 October 1989, 1.

8. The production and export history of this aircraft can be found in *Jane's All the World's Aircraft, 1986-87* (London: Jane's Publishing Company, 1986), 123-24.

9. For a recent assessment of arms export policy in the Federal Republic of Germany, see Michael Brzoska, "The Erosion of Restraint in West German Arms Transfer Policy," *Journal of Peace Research* 26 no. 2 (1989): 165-77.

10. For an excellent history of this transaction, see John Newhouse, "Politics and Weapons Sales," *New Yorker* (9 June 1986): 46-56.

11. Laurance and Mullen, op. cit., 78.

12. Louscher and Salomone, op. cit., 25.

13. Edward A. Kolodziej, *Making and Marketing Arms: The French Experience and Its Implications for the International System* (Princeton: Princeton University Press, 1987), 415.

14. Naoum Sloutsky, *The League of Nations and the Control of Trade in Arms.* Report No. C/74-8 (Cambridge: Center for International Studies, MIT, 1974), 116.

15. I first saw this term used in Ted Gurr, *Politimetrics* (Englewood Cliffs, N.J.: Prentice-Hall, Inc., 1972).

16. Michael Brzoska, "Arms Transfer Data Sources," *Journal of Conflict Resolution* 26 (March 1982): 87.

17. *World Military Expenditures and Arms Transfers 1988* (Washington, D.C.: U.S. Arms Control and Disarmament Agency, 1989), iii.

18. These changes came about as a result of paying more attention to Soviet arms transfers relative to other suppliers. They appear to result in more valid and reliable data on Soviet arms transfers. Interview with CIA official.

19. See Michael Brzoska and Thomas Ohlson, eds., *Arms Transfers to the Third World: 1971-1985* (Oxford: Oxford University Press, 1987), 365.

20. Kolodziej, op. cit.

21. U.S. Congressional Research Service, *Trends in Conventional Arms Transfers to the Third World by Major Suppliers, 1983-1990.* (Washington, D.C.: Library of Congress, 1991).

22. This is based on my experience as a consultant to a private firm that created an arms transfer event data base using periodicals, newspapers, wire services, and reference books on military affairs. For an account of the methodology and data sources, see Laurance and Mullen, op. cit., 92-96.

23. For an account of this transfer see Edward J. Laurance, "The New Gunrunning," *Orbis* (Spring 1989): 225-37.

24. Jay Mallin, Sr., "More Soviet Weapons Landed in Nicaragua," *Washington Times,* 5 June 1984, 1.

25. U.S. Department of Defense, *Soviet Military Power* (Washington, D.C.: GPO, 1985), 119-23.

26. Edward J. Laurance and Ronald G. Sherwin, "Understanding Arms Transfers Through Data Analysis," in *Arms Transfers to the Third World: The Military Buildup in Less Industrial Countries,* ed. Uri Ra'anan et al. (Boulder: Westview Press, 1978), 87-106; Michael Brzoska, "Arms Transfer Data Sources," *op. cit.*

27. *World Military Expenditures and Arms Transfers 1965-1974* (Washington, D.C.: U.S. Arms Control and Disarmament Agency, 1976), 5.

28. The SIPRI pricing system is explained in detail in Brzoska and Ohlson, op. cit., 352-59.

29. Brzoska and Ohlson, op. cit., 366.

30. *World Military Expenditures and Arms Transfers 1988,* op. cit., 130.

31. Brzoska, "Arms Transfer Data Sources," op. cit., 93.

32. For a thorough treatment of "price and availability" and other types of data generated by required reports, see U.S. Congress, Subcommittee on Arms Control, International Security and Science, Committee on Foreign Affairs, House, *U.S. Military Sales and Assistance Programs: Laws, Regulations, and Procedures,* Committee Print (Washington: Library of Congress, Congressional Research Service, 1985).

33. There was a case in 1978 where a government witness, at a congressional hearing, released information on all conventional arms transfer requests turned down by the U.S. government between January 1 and September 1, 1978. See *Aerospace Daily,* 12 December 1978.

34. "On the Eve: A Survey of West Germany," *Economist,* 28 October 1989.

35. In Brzoska's seminal review of arms transfer data, he uses the phrase "politically motivated sloppiness." Brzoska, "Arms Transfer Data Sources," op. cit., 102.

36. *World Military Expenditures and Arms Transfers 1988,* 133.

37. Edward A. Kolodziej, "Measuring French Arms Transfers: A Problem of Sources and Some Sources of Problems with ACDA Data," *Journal of Conflict Resolution,* 23 (June 1979): 195-227.

38. In Kolodziej's book, he gives concrete examples of all of these problems that occurred during his research. Kolodziej, *Making and Marketing Arms,* op. cit., 413-14.

39. Craig Etcheson, *Arms Race Theory* (Westport, Conn.: Greenwood Press, 1989).

40. Hans Rattinger, "From War to War: Arms Races in the Middle East," *International Studies Quarterly*, 20 (December 1976): 502.

41. Foreign Military Sales (FMS) agreements in current dollars, from DSAA data.

42. For a review of these approaches, see Ronald G. Sherwin and Edward J. Laurance, "Arms Transfers and Military Capability: Measuring and Evaluating Conventional Arms Transfers," *International Studies Quarterly*, 23, no. 1 (September 1979): 360–89.

43. The SIPRI pricing system is explained in detail in Brzoska and Ohlson, op. cit., 352–59.

44. Sherwin and Laurance, op. cit., 377–87.

45. The results reported in this research were recently used as a basis for ranking Third World navies in Michael A. Morris, *Expansion of Third-World Navies* (New York: St. Martins Press, 1987).

46. John H. Cushman, Jr., "The Stinger Missile: Helping to Change the Course of a War," *The New York Times,* 17 January 1988, 24; Oliver Roy, *The Lessons of the Soviet/Afghan War,* Adelphi Papers no. 259 (London: International Institute for Strategic Studies, Summer 1991).

47. Alexander George, "Case Studies and Theory Development: The Method of Structured, Focused Comparison," in *Diplomatic History: New Approaches,* ed. Paul Gorden Lauren (New York: Free Press, 1979), 43–68.

48. Klare estimates that the amount of illegal trade in the 1980s was $10 billion per year. See Michael T. Klare, "The State of the Trade: Global Arms Transfer Patterns in the 1980s," *Journal of International Affairs* 40 (Summer 1986), 1–22. Also see Laurance, "The New Gunrunning," op. cit.

49. Stephanie G. Neuman, "Offsets in the International Arms Market," in *World Military Expenditures and Arms Transfers 1985* (Washington, D.C.: U.S. Arms Control and Disarmament Agency, 1986), 35.

50. For a summary and evaluation of these reports on U.S. offset data, see U.S. General Accounting Office, *Military Exports: Analysis of an Interagency Study on Trade Offsets,* Report No. GAO/NSIAD-86-99BR (Washington, D.C.: GPO, 1988).

51. Robert Pear, "U.S. Ranked No. 1 in Weapons Sales," *New York Times,* 11 August 1991, 8.

52. "Arms Report Misleading, U.S. Claims," *Herald* (AP), 13 August 1991, 1.

# Chapter 3: International Arms Trade as Systems and Regimes

1. Ernst-Otto Cziempel and James Rosenau, *Global Changes and Theoretical Challenges* (Lexington, Mass.: Lexington Books, 1989), 2.

2. South Korea later canceled the F/A-18 deal in favor of the F-16, but the technology transfer provisions are roughly the same.

3. Cziempel and Rosenau, op. cit., 15.

4. Ibid., 16.

5. Kenneth N. Waltz, *Theory of International Politics* (Reading, Mass.: Addison-Wesley, Publishing Company, 1979), 57.
6. Ibid., 101.
7. Ibid., 123
8. K. J. Holsti, *International Politics* (Englewood Cliffs: Prentice-Hall, 1988).
9. Holsti, op. cit., 24.
10. Robert E. Harkavy, *The Arms Trade and International Systems* (Cambridge: Ballinger Publishing Company, 1975).
11. Ibid., 41–47.
12. Holsti, op. cit., 24.
13. Ibid., 447.
14. In Stephen D. Krasner's seminal book, *International Regimes* (Ithaca: Cornell University Press, 1983), the chapters by Krasner, Donald J. Puchala and Raymond F. Hopkins, and Ernst B. Haas all subscribe to this view.
15. Krasner, op. cit., 2.
16. Ibid.
17. Robert Jervis, "Security Regimes," in Krasner, op. cit., 188.
18. Puchala and Hopkins, in Krasner, op. cit., 62.
19. Dan E. Caldwell, *American-Soviet Relations* (Westport, Conn.: Greenwood Press, 1981), 28.
20. Ibid., 42.
21. David W. Tarr, *Nuclear Deterrence and International Security: Alternative Nuclear Regimes* (New York: Longman Publishing Group, 1991); Joseph S. Nye, Jr., "Nuclear Learning and U.S.-Soviet Security Regimes," *International Organization* 41, no. 3 (Summer 1987): 371–402.
22. Beverly Crawford and Stefanie Lenway, "Decision Modes and International Regime Change: Western Collaboration on East-West Trade." *World Politics* 37, no. 3 (April 1985): 375–403.
23. Ibid., 385.
24. What follows is a brief summary of John Simpson, "The Nuclear Non-proliferation Regime as a Model for Conventional Armament Restraint," in *Arms Transfer Limitations and Third World Security,* ed. Thomas Ohlson (New York: Oxford University Press, 1988), 227–40.
25. Krasner, op. cit., 2.

## Chapter 4: The Interwar Arms System

1. Robert E. Harkavy, *The Arms Trade and International Systems* (Cambridge: Ballinger Publisher Company, 1975).
2. Harkavy, op. cit., 10.
3. Harkavy, op. cit., 33.
4. Harkavy, op. cit., 41.
5. Ibid., 41–46.
6. Ibid., 46.
7. Ibid., 192.

8. The figures in table 4-2 must be seen in the light of the basic lack of control that national governments had over armaments manufacturers and traders. Numbers of aircraft are those exported by firms, not countries.

9. Robert E. Harkavy, "Arms and Political Influence: The Temporal Dimension" (Paper presented to the annual meeting of the International Studies Association, Toronto, February 1976), 59.

10. Harkavy, *Arms Trade,* 173.

11. Ibid., 79.

12. Harkavy designates Germany, Italy, Japan, and the USSR as "revisionist."

13. Harkavy, *Arms Trade,* table 3-15, pp. 84-85.

14. Ibid., 202-3.

15. Ibid., 126-27

16. For an assessment that utilizes the firm as an actor, see Mary Kaldor, "Vickers," in *The Baroque Arsenal* (New York: Hill and Wang, 1981), 29-53. Also see Basil Collier, *Arms and the Men: The Arms Trade and Governments* (London: Hamish Hamilton, 1980).

17. See notes 42 to 45, chapter 2.

18. Harkavy, *Arms Trade,* 44-45.

19. Ibid., 93.

20. Ibid., 151.

21. Ibid., 161.

22. Ibid., 170.

23. For example, archives of the British Vickers Company are available.

24. John E. Wiltz, *In Search of Peace: The Senate Munitions Inquiry, 1934-36* (Baton Rouge: Louisiana State University Press, 1963), 22.

25. Ibid., 231-32.

26. Ibid., 22.

27. Ibid., 23.

28. Wiltz makes the very interesting point that the Nye committee failed to exploit this issue. Did the nature of munitions impose upon the industry more exacting standards of conduct? "Surprisingly, in view of the comprehensiveness of its study, the committee failed to pursue this point, a major tactical error. The committee sought stringent control of the arms trade, and the distinctive character of munitions was the strongest argument for such restriction. Concentrating on this point the committee might have built a formidable case. Instead it banked its hopes on proving all the particulars of the pacifist indictment of the munitions trade." Ibid., 83-84.

29. Andrew J. Pierre, *The Global Politics of Arms Sales* (Princeton: Princeton University Press, 1982).

30. Robert E. Harkavy, "The New Geopolitics: Arms Transfers and the Major Powers' Competition for Overseas Bases," in *Arms Transfers in the Modern World,* ed. Stephanie G. Neuman and Robert E. Harkavy (New York: Praeger Publishers, 1979), 134.

31. Edward A. Kolodziej and Frederick S. Pearson. *The Political Economy of Making and Marketing Arms: A Test for the Systemic Imperatives of Order and Welfare,* Occasional Paper No. 8904 (University of Missouri-St. Louis: Center for International Studies, April 1989), 6.

## Chapter 5: The Evolution of Arms Trade Systems in the Postwar Period

1. Robert E. Harkavy, *The Arms Trade and International Systems* (Cambridge: Ballinger Publishing Company, 1975), p. xiv.
2. Cindy Cannizzo, "Trends in Twentieth-Century Arms Transfers," in *The Gun Merchants: Politics and Policies of the Major Arms Suppliers*, ed. Cindy Cannizzo (Elmsford, N.Y.: Pergamon Press, 1980), 3.
3. Uri Ra'anan et al., eds., *Arms Transfers to the Third World: The Military Buildup in Less Industrial Countries* (Boulder: Westview Press, 1978), xiii.
4. Stephanie G. Neuman and Robert E. Harkavy, eds., *Arms Transfers in the Modern World* (New York: Praeger Publishers, 1979).
5. Andrew J. Pierre, *The Global Politics of Arms Sales* (Princeton: Princeton University Press, 1982).
6. Michael Brzoska and Thomas Ohlson, eds., *Arms Transfers to the Third World: 1971-85* (Oxford: Oxford University Press, 1987).
7. For example, he did not use any of the ACDA data, which were not available until after his research was completed.
8. Harkavy, op. cit., 10.
9. Ibid., 93. See table 4-6.
10. Ibid., 193.
11. Henry Kuss, speech to American Ordnance Association, cited in Stockholm International Peace Research Institute, *The Arms Trade with the Third World* (New York: Humanities Press, 1971), 135. For more thorough treatments of this phenomenon, see John Stanley and Maurice Pearton, *The International Trade in Arms* (New York: Praeger Publishers, 1972).
12. USSR "industrialized" recipients included China, the Warsaw Pact nations, and Finland. For the United States they included NATO, Japan, and Australia. Lewis A. Frank, *The Arms Trade in International Relations* (New York: Praeger Publishers, 1969), 86-87, 102-3.
13. Michael D. Salomone and David J. Louscher, "Foreign Policy Priorities and Arms-Transfer Trends," *Crossroads* 15 (1985): 47.
14. An excellent summary of this period can be found in SIPRI, op. cit., 180-214.
15. Alden F. Mullins, Jr., *Born Arming: Development and Military Power in New States* (Stanford: Stanford University Press, 1987), 15.
16. Harkavy, op. cit., 61-76.
17. Christian Catrina, *Arms Transfers and Dependence* (New York: Taylor and Francis, 1988), Fig. 3-1, p. 46.
18. Brzoska and Ohlson, op. cit., app. 6-A, pp. 332-33.
19. U.S. Arms Control and Disarmament Agency, *World Military Expenditures and Arms Transfers 1963-73* (Washington: GPO, 1974).
20. Harkavy, op. cit., 116. Harkavy defines a *sole-supplier* relationship as one where a single donor has supplied all of the weapons received by a given recipient. A *dominant* relationship is one where a single donor has supplied 60 percent or more, on average, of all weapons systems or is a primary supplier in most all of them. This 100 percent/60 percent definition is maintained throughout this book.
21. One of the best treatments of this phenomenon remains George Thayer's *The War Business* (London: Paladin Press, 1970), 274-76.

22. Harkavy, op. cit., 156–61.

23. Paul Jabbar, *Not by War Alone: Security and Arms Control in the Middle East* (Berkeley: University of California Press, 1981).

24. SIPRI, op. cit., 101–3.

25. It should be noted that despite being capable of major-supplier status in terms of equipment available, it was not until 1955 that the USSR made the foreign policy decision to supply other than the most orthodox communist states.

26. For a survey of global attitudes of the nuclear military balance, see Donald C. Daniel, ed., *International Perceptions of the Superpower Military Balance* (New York: Praeger Publishers, 1983).

27. *SIPRI Yearbook 1981* (London: Taylor and Francis, 1982), 70, citing RAND report. Also see Michael Mihalka, "Supplier-Client Patterns in Arms Transfers: The Developing Countries," in *Arms Transfers in the Modern World*, ed. Stephanie G. Neuman and Robert E. Harkavy (New York: Praeger Publishers, 1979), 56–57.

28. Pierre, op. cit., 3.

29. It is ironic that at the same time as arms firms were becoming multinational for the first time since the interwar years, nonmilitary multinational firms had been playing a dominant role in the international economy and were the target of a great deal of international criticism.

30. *Jane's All the World's Aircraft, 1975–76* (London: Jane's Publishing Company, 1975).

31. "Third World Arms Production," in *World Military Expenditures and Arms Transfers 1969–78* (Washington: U.S. Arms Control and Disarmament Agency, 1979).

32. Stephanie G. Neuman, "International Stratification and Third World Military Industries." *International Organization*, 38, no. 1 (Winter 1984): 197.

33. Ibid., 172–73.

34. *SIPRI Yearbook 1981*

35. Brzoska and Ohlson, op. cit., 8.

36. Stephanie Neuman, "International Stratification and Third World Military Industries," *International Organization*, 38, no. 1 (Winter 1984): 167–97.

37. For a complete list of sole-, predominant-, and multiple-supplier relationships in this period, by client and supplier, see Michael Mihalka, "Supplier-Client Patterns," op. cit.

38. Catrina, op. cit., 28.

39. Despite a decline in the market shares of the United States and the USSR during this period, their actions continued to mirror systemic conditions.

40. George E. Hudson, "The Economics of Soviet Arms Transfers," in *Marketing Security Assistance*, ed. David J. Louscher and Michael D. Salomone (Lexington, Mass.: Lexington Books, 1987), 71.

41. Brzoska and Ohlson, op. cit., 9.

42. William B. Quandt, "Influence through Arms Supply: The U.S. Experience in the Middle East," in *Arms Transfers to the Third World: The Military Buildup in Less Industrialized Countries*, ed. Uri Ra'anan et al. (Boulder, Colo.: Westview Press, 1978), 121–30.

43. Stephanie G. Neuman, *Military Assistance in Recent Wars: The Dominance of the Superpowers* (New York: Praeger Publishers, 1986), 126.

44. Stephanie G. Neuman, "Arms, Aid and the Superpowers," *Foreign Affairs* (Summer 1988): 1044–65.

45. Quandt, op. cit.
46. Stephanie G. Neuman, *Military Assistance*. See also Rodney W. Jones and Steven A. Hildreth, *Modern Weapons and Third World Powers* (Boulder, Colo.: Westview Press, 1984).
47. Pierre, op. cit.
48. Mihalka, op. cit., table 4–10, p. 73.
49. Bjorn Hagelin, "Multinational Weapon Projects and the International Arms Trade," in *SIPRI Yearbook 1984* (London: Taylor and Francis, 1985), 151–63.
50. Keith Krause, "Arms Transfers, External Powers and Conflict Management in the Persian Gulf" (Unpublished paper, York University, 1990), 12.
51. Ibid., 14.
52. The literature on this issue is extensive. For an empirical treatment that demonstrates that the effect of arms imports on developing economies varies by type of recipient, see Robert Looney, *Third-World Military Expenditures and Arms Production* (New York: St. Martin's Press, 1988). For an analysis that focuses on the deleterious effects of arms imports, see Nicole Ball, *Security and Economy in the Third World* (Princeton: Princeton University Press, 1988).
53. For a typical treatment of these negative consequences, see Steven J. Rosen, "The Proliferation of New Land-based Technologies: Implications for Local Military Balances," in *Arms Transfers in the Modern World,* ed. Stephanie G. Neuman and Robert E. Harkavy (New York: Praeger, 1979), 109–30.
54. Several postmortems of these talks have been written. For one written by a participant, see Barry M. Blechman et al., "Pushing Arms," *Foreign Policy* 46 (Spring 1982): 138–54.
55. For a recent and comprehensive treatment of arms trade control in this period, see Keith Krause, "Constructing Regional Security Regimes and the Control of Arms Transfers," *International Journal* 45 (Spring 1990): 386–423.
56. For a recent in-depth treatment of this case, see Signe Landgren, *Embargo Disimplemented: South Africa's Military Industry* (New York: Oxford University Press, 1989).

## Chapter 6: The Declining Bipolar Arms Trade System, 1980–92

1. *SIPRI Yearbook 1990* (New York: Oxford University Press, 1990), 207.
2. In the final chapter of this book, the post-1992 arms trade system is discussed in detail.
3. The Nunn Amendment programs were those created in 1986 legislation that provided special funds for the U.S. military services to use in acquiring weapons systems developed jointly with European industries, and to test current European systems already in use. The program was designed to foster cooperation in NATO and to eliminate redundant weapons systems within the alliance.
4. "Arms Production," in *SIPRI Yearbook 1990,* 325–43.
5. For a typical commentary on this reality, see "Last of the Old Time Summits," *New York Times,* 25 July 1991.
6. Richard Grimmett, *Trends in Conventional Arms Transfers to the Third*

*World by Major Supplier, 1982-1989* (Washington: Congressional Reference Service, 1990), 49.

7. Francis Fukuyama, "The End of History," *National Interest* 16 (Summer 1989): 3-18.

8. Rita Tullberg, "Military-related Debt in Non-Oil Developing Countries, 1972-82," in *SIPRI Yearbook 1985* (London: Taylor and Francis, 1985), 445-58; Somnath Sen, "Debt, Financial Flows and International Security," in *SIPRI Yearbook 1990,* 203-17.

9. *SIPRI Yearbook 1990,* 219-23.

10. Steven Greenhouse, "Rebuilding Troubled Regions Will Strain World's Finances," *New York Times,* 26 March 1991, A1.

11. Elie Kedourie, "Avoiding a Third Gulf War," *New York Times,* 13 March 1991.

12. "Prospects for Third World Brighten," *Financial Times,* 8 July 1991.

13. Stephen Kinzer, "Germany to Cut Aid to Countries That Spend Heavily on Weapons," *New York Times,* 3 August 1991, 3.

14. For the most recent summary of these assessments, see Somnath Sen, "Debt, Financial Flows and International Security," in *SIPRI Yearbook 1991* (Oxford: Oxford University Press, 1991), 181-95.

15. *Global Arms Trade: Commerce in Advanced Military Technology and Weapons* (Washington: U.S. Congress, Office of Technology Assessment, 1991), 35-37.

16. A good example would be the *SIPRI Yearbook 1990,* which includes for the first time a section on defense industries. This occurred as SIPRI recognized the increasing importance of defense firms as units of analysis in the international arms trade system. Interview with Ian Anthony and Herbert Wulf at SIPRI, April 1990.

17. Julie Johnson, "China Would Back Embargo on Iran," *New York Times,* 8 March 1987; Nicholas D. Kristof, "China Said to Plan New Arms Sales; U.S. Is Concerned," *New York Times,* 10 June 1991, 1; Timothy V. McCarthy, *A Chronology of PRC Missile Trade and Developments.* (Monterey: Monterey Institute of International Studies, International Missile Proliferation Project, 1992).

18. Muller, Harald, *The Controversy over West German Export Policy,* Research Note 1. (Monterey, Calif.: Monterey Institute of International Studies, 1990).

19. For a brief description of the phenomenon, see Gaylord Shaw and William C. Rempel, "Billion-Dollar Iran Arms Search Spans U.S. Globe," *Los Angeles Times,* 4 August 1985, 1.

20. Michael Brzoska, "Profiteering on the Iran-Iraq War," *Bulletin of the Atomic Scientists* (June 1987): 42-45; Michael T. Klare, "Secret Operatives, Clandestine Trades: The Thriving Black Market for Weapons," *Bulletin of the Atomic Scientists* (April 1988): 16-24; see also the chapter on the international trade in arms in *SIPRI Yearbook 1987* (Oxford: Oxford University Press, 1987).

21. "De Facto De-regulation of the Arms Trade." *International Defense Review* (April 1986): 397; "Some Rules of the Road for the Arms Traffic," *International Defense Review* (June 1987): 701.

22. For a thorough treatment of the impact of illegal trade on the United States, see Edward J. Laurance, "The New Gunrunning," *Orbis* (Spring 1989): 225-37.

23. Jonathan Beaty and S. C. Gwynne, "The Dirtiest Bank of All," *Time,* 29 July 1991, 42-47.

24. "New Sellers in Arms Bazaar: Superpowers Face Competition from Surprising Corners of the World." *U.S. News & World Report,* 3 February 1986, 37-39; Aaron

Klieman, *Israel's Global Reach: Arms Sales as Diplomacy* (New York: Pergamon-Brassey's International Defense Publishers, 1985); Stephanie G. Neuman, "Third World Arms Production and the Global Arms Transfer System," in *Arms Production in Developing Countries*, ed. James Everett Katz (Lexington, Mass.: Lexington Books, 1984); Joseph F. Clare, Jr., "Whither the Third World Arms Producers," in *World Military Expenditures and Arms Transfers 1986* (Washington: U.S. Arms Control and Disarmament Agency, 1987).

25. Chile's cluster bomb cost $500, compared to $2,000 for an identical item made in the United States.

26. *SIPRI Yearbook 1990*, 247–48.

27. The definitive work on the rise and fall of Brazilian arms exports is Scott D. Tollefson, "Brazilian Arms Transfers, Ballistic Missiles, and Foreign Policy" (Ph.D. diss., Johns Hopkins University, 1991). Also see James Brooke, "Peace Unhealthy for Brazilian Arms Industry," *New York Times*, 26 February 1990, A4.

28. *SIPRI Yearbook 1990*, 237–41. See also Gerald M. Steinberg, "The Impact of New Technology on the Arab-Israeli Military Balance," in *The Soviet-American Competition in the Middle East*, ed. Steven L. Speigel, Mark A. Heller, and J. Goldberg (Lexington, Mass.: Lexington Books, 1988.)

29. Michael Brzoska, "The Erosion of Restraint in West German Arms Transfer Policy," *Journal of Peace Research* 26, no. 2 (1989): 165–77.

30. *SIPRI Yearbook 1990*, 219.

31. *SIPRI Yearbook 1991*, 219.

32. Robert Pear, "U.S. Ranked No. 1 in Weapons Sales," *New York Times*, 11 August 1991, 8.

33. Data for second-tier suppliers in this figure include non-Soviet Warsaw Pact countries.

34. Charles Babcock, "Anti-arms Sales Drive Was 'Modestly Successful,'" *Washington Post*, 10 December 1986, p. 22; For an in-depth treatment of Operation Staunch, see Charles Roller and Dorothy Major, "Ramifications of Illegal U.S. Arms Exports" (Master's thesis, Naval Postgraduate School, 1989).

35. Michael Wines, "Third World Seeks Advanced Arms," *New York Times*, 26 March 1991, A6; Philip Finegan and Robert Holzer, "Third World Offers New Threat," *Defense News*, 24 June 1991, 20.

36. Michael Brzoska and Thomas Ohlson, eds., *Arms Transfers to the Third World: 1971–85* (Oxford: Oxford University Press, 1987), 12.

37. Michael A. Morris, *Expansion of Third-World Navies* (New York: St. Martin's Press, 1987).

38. Ibid., 101–2

39. Ibid., 104.

40. Ian Anthony, *The Naval Arms Trade* (Oxford: Oxford University Press, 1990), 5.

41. Ibid., 46–53.

42. Ibid., 55.

43. For an inventory of the maritime patrol craft acquired and jurisdictional claims by country, see Ian Anthony, "The Naval Arms Trade and Implications of Changes in Maritime Law," in *SIPRI Yearbook 1988* (Oxford: Oxford University Press, 1989), 270–273.

44. "Adm. Brooks Names Names," *Jane's Defence Weekly*, 16 March 1991, 366.

45. Glenn Charles Ackerman, "Responding to the Threat from Third World Air Defense Systems: A Comparison of U.S. Policy Options" (Masters thesis, Naval Postgraduate School, 1990).

46. "Iraq's 'SCUD' Programme—The Tip of the Iceberg," *Jane's Defence Weekly*, 2 March 1991, 301–3.

47. Israel is codeveloping with the United States a missile, the Arrow, that is designed to defeat a ballistic missile. Despite the success of the Patriot in the Gulf War, it was not designed to counter the type of surface-to-surface missile seen as a threat in the post-1995 period. In a new effort, a U.S.-European industrial alliance has proposed a set of joint programmes to develop technology that could combat the proliferation of ballistic missiles. See *Cooperative Strategies: High Technology Security Cooperation, A Transatlantic Industrial Perspective*. (Arlington, VA: U.S.-CREST, 1991); Paul Abrahams, "Arms Groups Seek Joint Missile Defence Project," *Financial Times*, 1 June 1991.

48. William C. Potter and Adam Stulberg, "The Soviet Union and the Spread of Ballistic Missiles," *Survival* 32, no. 6 (November/December 1990): 543–57.

49. For recent comprehensive studies, see Martin Navias, *Ballistic Missile Proliferation in the Third World*, Adelphi Paper no. 252 (London: International Institute for Strategic Studies, 1990); Janne Nolan, *Trappings of Power: Ballistic Missiles in the Third World* (Washington: Brookings Institution, 1991). A bibliography on this issue is included at the end of this book.

50. Ian Anthony et al., "The Trade in Major Conventional Weapons," *SIPRI Yearbook 1991*, 225–27.

51. For a recent example, see "Weapon System Upgrades: Defense Cuts Drive Market," *Aviation Week and Space Technology*, 22 July 1991, 38–58.

52. Arthur Alexander, "NATO Cooperation and the Export of Arms" (Unpublished paper presented to the Arms Trade Workshop, 27 February 1988).

53. Mark Kirk, "The Arms Trade No One Talks About," *Christian Science Monitor*, 14 September 1984, 14.

54. Barbara Starr, "Row Looms over CFE Surplus Sale," *Jane's Defence Weekly*, 18 May 1991, p. 813; Ian Anthony et al., "Trade in Major Conventional Weapons," *SIPRI Yearbook 1991*, 205–6.

55. Commercial arms exports from the United States must still go through a licensing process. However, as is true in most supplier states, this process is significantly less demanding from the perspective of control by the government.

56. Thomas L. Selden, "The Internationalization of Military Aircraft," *DISAM Journal* (Summer 1986): 106; see also Bjorn Hagelin, "Multinational Weapon Projects and the International Arms Trade," *SIPRI Yearbook 1984* (London: Taylor and Francis, 1984), 151–63.

57. For a comparison of three types of aircraft production—national (YF-22A), national control of production but internationalized subsystems (JAS-39), and multinational (EFA)—see *SIPRI Yearbook 1991*, 224.

58. Alexander, op. cit.

59. S. Faltas, *Arms Markets and Armament Policy* (Dordrecht: Martinus Nijhoff Publishers, 1986).

60. The most recent U.K. government report on international collaboration cited 15 percent of the procurement by the Ministry of Defense as non-British, including programs with the United States.

61. Andrew Moravcsik, "The European Armaments Industry at the Crossroads," *Survival* 32, no. 1 (January/February 1990): 65-85.

62. For an introduction to this trend, see Moravcsik, op. cit.; Martyn Bittleston, *Cooperation or Competition? Defence Procurement Options for the 1990s,* Adelphi Papers no. 250 (London: International Institute for Strategic Studies, 1990); Margaret Blunden, "Collaboration and Competition in European Weapons Procurement: The Issue of Democratic Accountability," *Defense Analysis* 5, no. 4 (1989): 294.

63. John G. Roos, "Allied Procurement, R&D Funding Head for Tailslide in '92," *Armed Forces Journal International,* March 1991, 24.

64. For an in-depth treatment of these developments and further data on European multinational production, see Edward J. Laurance, "The Future of Arms Production and Export in Western Europe: A Model and Some Preliminary Indications" (Unpublished paper presented to the International Studies Association, March 1991).

65. Stephanie G. Neuman, "Coproduction, Barter and Countertrade: Offsets in the International Arms Market," *Orbis* (Spring 1985): 183-213.

66. Ingemar Dorfer, *Arms Deal: The Selling of the F-16* (New York: Praeger Publishers, 1983).

67. *Third Annual Report on the Impact of Offsets in Defense-Related Exports* (Washington: Office of Management and Budget, 1987), 5-24.

68. Brzoska and Ohlson, op. cit., 131.

69. Richard G. O'Lone, "Boeing, Saudi Arabia Tie Defense System to Economic Package," *Aviation Week and Space Technology,* 3 June 1985, 155-58.

70. Tracy E. DeCourcy, "Countertrade and the Arms Trade in the 1980s," in *Marketing Security Assistance: New Perspectives on Arms Sales,* ed. David J. Louscher and Michael D. Salamone (Lexington, Mass.: Lexington Books, 1987), 167.

71. Ibid., 170.

72. Ibid., 172.

73. See Michael T. Klare, "The Unnoticed Arms Trade: Exports of Conventional Arms-Making Technology," *International Security* (Fall 1983): 68-90; and David J. Louscher and Michael D. Salomone, *Technology Transfer and U.S. Security Assistance: The Impact of Licensed Production* (Boulder, Colo.: Westview Press, 1986).

74. *Global Arms Trade: Commerce in Advanced Military Technology and Weapons* (Washington: U.S. Congress, Office of Technology Assessment, June 1991), 13.

75. "Exhibition Calendar," *International Defense Review,* September 1987, 1281.

76. Michael T. Klare, "The State of the Trade: Global Arms Transfer Patterns in the 1980s," *Journal of International Affairs* (Summer 1986): 14.

77. Aaron Karp, "The Trade in Conventional Weapons," *SIPRI Yearbook 1988,* 190.

78. *SIPRI Yearbook 1990,* 246.

79. Table derived from data gathered by author.

80. *SIPRI Yearbook 1988,* 190.

81. Jeffrey M. Lenorovitz, "COCOM Eases Restrictions on Export of High Technology Equipment to Eastern Bloc," *Aviation Week and Space Technology,* 10 June 1991, 73; for an in-depth and current treatment of these controls and their future, see *Finding Common Ground: U.S. Export Controls in a Changed Global Environment,* Report by the Committee on Science, Engineering and Public Policy: National Academy of Sciences (Washington: National Academy Press, 1991).

82. For another analytical treatment of the problem, see Robert E. Harkavy, "Arms Resupply during Conflict: A Framework for Analysis," in *The Economics of Military Expenditures*, ed. Christian Schmidt (London: Macmillan Press, 1987), 239–83.

83. See Stephanie G. Neuman, *Military Assistance in Recent Wars: The Dominance of the Superpowers* (New York: Praeger Publishers, 1986). It should be noted in this discussion that national actors are reluctant to accept such conclusions, as I am sure would be revealed if we were talking with the commanding general of the British forces that prevailed in this conflict. However, the fact remains that despite the skill and courage of the British Harrier pilots, an around-the-clock U.S. resupply of AIM-9L Sidewinder air-to-air missiles played no small part in their success.

84. Mark Fineman, "Afghan Pilots Say They've Taken Sting Out of Stinger," *Los Angeles Times*, 2 May 1989.

85. "Sri Lanka: Not All the Guns Were Handed In," *Economist*, 19 September 1987, 44–45.

86. "Southern Yemen Cites Death of 4,230 in January Fighting," *New York Times*, 7 December 1986.

87. A quick re-read of Orwell's *1984* will reveal that this is exactly what the three superpowers of his world concluded after a brief nuclear exchange produced negative effects on their home territories.

88. "Beirut Seeks Full PLO Withdrawal," *New York Times*, 9 July 1991.

89. Thalif Deen, "USA and Soviets to End Arms Aid," *Jane's Defence Weekly*, 15 June 1991, 1001.

90. "Dark Skies over Cambodia," *Economist*, 23 June 1991.

91. "Defense News Roundtable: Industry Mulls Arms Sales after Gulf War," *Defense News*, 15 July 1991, 8.

92. This also was the conclusion (Finding 6) of the most recent Office of Technology Assessment study, *Global Arms Trade: Commerce in Advanced Military Technology and Weapons* (Washington: U.S. Congress, Office of Technology Assessment, 1991), 16.

93. For a most recent example, see Maureen Dowd, "Bush, in Athens, Sets New Arms Aid," *New York Times*, 20 July 1991, A1.

94. Krause first discussed this concept of structural influence in "Conflict Management, Arms Transfers, and the Arab-Israeli Conflict," in *Conflict Management in the Middle East*, ed. Gabriel Ben-Dor and David B. Dewitt, (Lexington, Mass.: Lexington Books, 1987), 209–38.

95. Keith Krause, "Arms Transfers, External Powers and Conflict Management in the Persian Gulf" (Unpublished paper, York University, 1990), 15.

96. *SIPRI Yearbook 1991*, 202–3.

97. John W. Lewis, Hua Di, and Xue Lita, "Beijing's Defense Establishment: Solving the Arms-Export Dilemma," *International Security* 15, no. 4 (Spring 1991): 87–109; Gary Milhollin and Gerard White, *Bombs from Beijing: A Report on China's Nuclear and Missile Exports* (Washington: Wisconsin Project on Nuclear Arms Control, 1991).

98. "Western nations have helped arm Iraq, the rest of the Middle East, and other regions with little concern or oversight about the near- or far-term consequences. . . . "As the Islamic revolution in Iran has shown, once transferred, modern weapons can outlast the governments they were intended to support. As the war with Iraq has shown,

arms may outlast the good will of the leaders to whom they were supplied. Highly armed adversaries make it more difficult for the United States to protect its interests, increasingly so in the future if the United States stays its post-Cold War course of reducing its armed forces and defense expenditures." *Global Arms Trade,* op. cit., 1, 17-18.

99. Janne Nolan, "The Global Arms Market after the Gulf War: Prospects for Control," *Washington Quarterly,* (Summer 1991): 125.

100. See "Kuwait: How the West Blundered," *Economist,* 29 September 1990; Murray Waas, "Who Lost Kuwait?," *Village Voice,* 22 January 1991.

101. Alan Cowell, "Egypt Proposes Regional Arms Control Plan," *New York Times,* 5 July 1991, A6.

102. Robert Mauthner, "A Question of Confidence," *Financial Times,* 5 June 1991, 19.

103. Nolan, op. cit. 126.

104. Seth Carus, *Ballistic Missiles in the Third World,* CSIS Washington Paper 146 (New York: Praeger Publishers, 1990), 53-66.

105. Ronald van de Krol, "U.K. Eases Syria Arms Stance," *Financial Times,* 11 July 1991, 4.

106. "Slovenia Agrees to Talks over Borders," *Financial Times,* 6-7 June 1991, 22.

## Chapter 7: Explaining National Arms Transfer Behavior and System Transformation at the Systemic Level

1. For but one example, see Christian Catrina, *Arms Transfers and Dependence* (New York: Taylor and Francis, 1988), 69-76.

2. See previous data on USSR hard currency sales (table 5-15).

3. *Jane's Defence Weekly,* 22 September 1990.

4. "Arms for Sale," *Newsweek,* 8 April 1991, 22-29; "Seeking New Customers," *Aviation Week and Space Technology,* 22 April 1991, 13.

5. Nick Cook, "Soviets Offer Yak-141 to India," *Jane's Defence Weekly,* 29 June 1991, 1164-65.

6. Alvin Rubenstein, *Red Star on the Nile* (Princeton: Princeton University Press, 1977).

7. John David Carlson, *Soviet Naval Arms Transfers 1945-1985: Domestic Factors and Constraints* (Master's thesis, Naval Postgraduate School, 1985).

8. Paul Ferrari et al., *U.S. Arms Exports: Policies and Contractors* (Cambridge: Ballinger Publishing Company, 1988), 37-64.

9. *Global Arms Trade: Commerce in Advanced Military Technology and Weapons* (Washington: U.S. Congress, Office of Technology Assessment, 1991), 1, 13.

10. "USA Defence Exporter: Finding a New Balance," *Jane's Defence Weekly,* 18 May 1991, 831-44.

11. These actions are summarized in *Global Arms Trade,* op. cit.; also see Steven Greenhouse, "Selling Planes That Won the War," *New York Times,* 21 June 1991, C1.

12. "USA Defence Exporter," op. cit.; Walter S. Mossberg and Rick Wartzman, "Back to the Race: Mideast Arms Outlays Seem Unlikely to Face Any Tough New Curbs," *Wall Street Journal,* 4 March 1991, A1; Patrick E. Tyler, "As the Dust Settles,

Attention Turns to New Arms Sales," *New York Times,* 24 March 1991, E3; "Mr. Bush Waffles on Mideast Arms," *New York Times,* 31 May 1991.

13. David Silverberg, "House Panel Kills Lending Plan," *Defense News,* 3 June 1991, 3.

14. "Israel Wants U.S. to Increase Military Aid," *Financial Times,* 5 July 1991, 4.

15. "U.S. Aid Decision Upsets Pakistanis," *New York Times,* 16 October 1990.

16. Keith Bradsher, "Baker Warns China against Selling New Missiles," *The New York Times,* 13 June 1991, A3.

17. U.S. Office of Management and Budget, *Offsets in Military Exports* (Washington: Executive Office of the President, 1989).

18. The response to this requirement is found in *Offsets in Military Exports* (Washington: Office of Management and Budget, 1990).

19. Remarks of R. Lee Hessler, Business Manager, Information Systems Division, Boeing Company. Reported in "The Impact of Offsets on Defense Related Exports," *DISAM Journal* 11, no. 1 (Fall 1988): 86–91.

20. For the latest book, see James Adams, *Engines of War: Merchants of Death and the New Arms Race* (New York: Atlantic Monthly Press, 1990).

21. For further empirical data, see Edward J. Laurance, "The New Gunrunning," *Orbis* 33, no. 2 (Spring 1989): 225–37.

22. George J. Church, "The Murky World of Weapons Dealers," *Time,* 19 January 1987, 26; Elaine Sciolino, "The Arms Market: Shadowy World," *New York Times,* 4 December 1986, 17; Martin Tolchin, "U.S. Policy 'Privatized' In Iran Deal, Critics Say," *New York Times,* 20 January 1987, 5; Jeff Gerth, "Arms Dealers Linked to U.S. Policy Shift," *New York Times,* 16 January 1987, 4; Stephen Engelberg, "From an Iranian Middleman in the Weapons Dealings, His Side of the Story," *New York Times,* 22 June 1987, 6.

23. A detailed examination of this phenomenon would illuminate the extent of the impact on the illegal arms trade, but such an examination is beyond the scope of this book. For a recent treatment, see William Branigan, "Mexico Cracks Major Arms, Drug-Trafficking Ring," *Washington Post,* 26 February 1988, A25; Larry Rohter, "From Brazil to Peru to Jamaica, Gun Smugglers Flock to Florida," *New York Times,* 11 August 1991, 1; Peter C. Unsinger and Harry W. Moore, eds., *The International Legal and Illegal Trafficking of Arms* (Springfield: Charles C Thomas Publisher, 1989).

24. Philip Shenon, "U.S. Accuses 2 Egyptian Colonels in Plot to Smuggle Missile Material," *New York Times,* 25 June 1988, 1.

25. James Brooke, "Peace Unhealthy for Brazilian Arms Industry," *New York Times,* 26 February 1990, A4; Thomas Kamm, "War Levels Brazil's Defense Firms, Which Thrived on Iraq's Purchases," *Wall Street Journal,* 5 February 1991, A14; "Industry after Eldorado," *Jane's Defence Weekly,* 15 June 1991, 1037, 1040.

26. Scott D. Tollefson, "Brazilian Arms Transfers, Ballistic Missiles, and Foreign Policy" (Ph.D. diss., Johns Hopkins University; 1991).

27. Rene Luria, "The Argentine Military and Industry Confront the Crisis," *International Defense Review* 6 (1990): 659–61; "Israeli Industry Update," *Jane's Defence Weekly,* 8 June 1991, 975–81.

28. "Arms Windfall Dilemma for Turkey," *Financial Times,* 27 June 1991, 6.

29. See Michael Brzoska and Thomas Ohlson, eds., *Arms Transfers to the Third World: 1971–85* (Oxford: Oxford University Press, 1987), app. 4A, 4B, and 7.

30. John W. Lewis, Hua Di, and Xue Lita, "Beijing's Defense Establishment: Solving the Arms-Export Dilemma," *International Security*, 15, no. 4 (Spring 1991): 87–109; Gary Milhollin and Gerard White, *Bombs from Beijing: A Report on China's Nuclear and Missile Exports* (Washington: Wisconsin Project on Nuclear Arms Control, 1991).

31. Robert Pear, "U.S. Ranked No. 1 in Weapons Sales," *New York Times*, 11 August 1991, 8.

32. For a thorough treatment of these shifts, see Anne Gilks and Gerald Segal, *China and the Arms Trade* (New York: St. Martin's Press, 1985).

33. Craig R. Whitney, "Arms Exports to End, Czech Foreign Minister Says," *New York Times*, 25 January 1990, A1.

34. John Tagliabue, "Scrap the Tanks? For Slovak Town, It's Call to Arms," *New York Times*, 4 March 1991, A4; George Leopold, "Czechs Try to End Arms Exports in Weak Economy," *Defense News*, 17 June 1991, 8.

35. Martin Navias, *Ballistic Missile Proliferation in the Third World*, Adelphi Papers no. 252 (London: International Institute for Strategic Studies, 1990).

36. "Minister Helped British Firms to Arm Saddam," *Sunday Times*, 2 December 1990, 1.

37. Jill Abramson and Edward T. Pound, "If Crisis Eases, Iraq Would Still Pose a Threat For Which the U.S. Must Shoulder Some Blame," *Wall Street Journal*, 7 December 1990.

38. "Frontline: The Arming of Iraq." Public Broadcasting System, 11 September 1990. Tape available from Films Incorporated, Chicago, Illinois.

39. Proposed amendment to the Foreign Military Sales Act, Section 1, Chapter 1, 17 December 1973. Senate Foreign Relations Committee.

40. Rick Wartzman, "Lockheed Wins South Korean Order for Planes," *Wall Street Journal*, 11 December 1990, A2.

41. "The JDW Interview," *Jane's Defence Weekly*, 1 June 1991, 936.

42. The most recent security assistance testimony is particularly noteworthy for its lack of emphasis on the influence rationale in the wake of the decline in East-West tensions.

43. "Seeking a New World," *Los Angeles Times*, 11 December 1990, H2–3.

44. Stephen J. Genco, "Integration Theory and System Change in Western Europe: The Neglected Role of Systems Transformation Episodes," in Ole R. Holsti, Randolph M. Siverson, and Alexander George, *Change in the International System* (Boulder, Colo. Westview Press, 1980), 68.

45. Ian Anthony, Agnes Courades Allebeck, Paola Miggiano, Elizabeth Skons, and Herbert Wulf, "The Trade in Major Conventional Weapons," in *SIPRI Yearbook 1992: World Armaments and Disarmament* (Oxford: Oxford University Press, 1992), Chapter Eight.

46. James Rosenau, *Turbulence in World Politics* (Princeton: Princeton University Press, 1990), 72.

47. Ibid., 76.

48. Edward A. Kolodziej and Frederick S. Pearson, *The Political Economy of Making and Marketing Arms: A Test for the Systemic Imperatives of Order and Welfare*, Occasional Paper No. 8904 (University of Missouri-St. Louis: Center for International Studies, 1989), 14.

49. Ibid., 30.

50. Ibid., 35.
51. Ibid., 40.
52. Keith Krause, "The Political Economy of the International Arms Transfer System: The Diffusion of Military Technique via Arms Transfers," *International Journal* 45 (Summer 1990): 689.
53. Ibid., 691-96.
54. Ibid., 700.
55. Ibid., 700-701.
56. Stephanie G. Neuman, "International Stratification and Third World Military Industries," *International Organization* 38, no. 1 (Winter 1984): 167-97.
57. Jill Abramson and Edward T. Pound, "If Crisis Eases, Iraq Would Still Pose a Threat For Which the U.S. Must Shoulder Some Blame," *Wall Street Journal*, 7 December 1990, A16.
58. Margaret Blunden, "Collaboration and Competition in European Weapons Procurement: The Issue of Democratic Accountability," *Defense Analysis* 5 no. 4 (1989): 291-304; Frederic S. Pearson, "Political Change and World Arms Export Markets: Impact on the Structure of West European Arms Industries," in Michael Brzoska and Peter Lock, eds. *Restructuring of Arms Production in Western Europe* (Oxford: Oxford University Press, 1992), 56-58.
59. See "Third World Arms Industries: Swords not Ploughshares," *Economist*, 23 March 1991, 50-51.
60. William J. Broad, "Soviet Woes Tarnish Once-Shining Space Efforts," *New York Times*, 8 January 1991, B5; William J. Broad, "Russian Seeks a U.S. Buyer for the World's Biggest Rocket," *New York Times*, 9 July 1991, B5.
61. Elisabeth Rubinfien, "Soviet Military Is Shaken by Allies' Triumph over Its Former Protege," *Wall Street Journal*, 4 March 1991, A8; Gabriel Schoenfeld, "The Loser of the Gulf War Is . . . the Soviet Military," *Wall Street Journal*, 19 March 1991, A24.
62. For one example, see "Call to Rethink Foreign Military Assistance Program," *Krasnaya Zvezda*, 3 October 1990, 1. Translation in *JPRS-UMA-90-028* (Washington: Foreign Broadcast Information Service, 17 December 1990), 81-82; also see Flora Lewis, "Here We Go Again—Arming the Middle East," *New York Times*, 21 March 1991, A19.
63. As one example, in November 1991 a manager of a missile factory in Siberia made a presentation/sales pitch at the Institute for East-West Security Studies in New York. Interview with research associate at the Institute for East-West Security Studies.
64. The literature on conversion in the former Soviet Union is extensive. A two-volume series was produced from a conference held in Moscow in 1990. *Conversion: Economic Adjustments in an Era of Reductions.* (New York: United Nations, 1991). For a thorough treatment see Julian Cooper, *The Soviet Defence Industry: Conversion and Reform* (London: Pinter Publishers, 1991).
65. Eric Schmitt, "U.S. Worries About Spread of Arms From Soviet Sales," *New York Times*, 16 November 1991.
66. William J. Broad, "A Soviet Company Offers Nuclear Blasts for Sale to Anyone With Cash," *New York Times*, 7 November 1991; William C. Potter, "Russia's Nuclear Entrepreneurs," *ibid*.
67. Author's observations and interviews with participants in the panel developing the U.N. arms trade register, January and April 1992.

68. David Silverberg, "Commerce Halts MoU Exchange on Fighter Pact with South Korea," *Defense News*, 22 January 1990, 4; Damon Darlin, "Seoul Tells U.S. It May Stall $3 Billion Jet-Fighter Deal," *Wall Street Journal*, 10 April 1990, 7; David E. Sanger, "General Dynamics Gets Korea Jet Deal," *New York Times*, 29 March 1991, C1.

69. U.S. Office of Management and Budget, *Offsets in Military Exports* (Washington: Executive Office of the President, 1989).

70. Steven Pearlstein, "Hard Sell for U.S. Arms," *Washington Post*, 7 April 1991, H1.

71. The reporting on this development was extensive, much of it in German. One of the most in-depth treatments in English is Kenneth R. Timmerman, *The Poison Gas Connection: Western Suppliers of Unconventional Weapons and Technologies to Iraq and Libya* (Los Angeles: Simon Wiesenthal Center, 1990). The best source in German is Hans Leyendecker and Richard Rickelmann, *Exporteure Des Todes: Deutscher Rustungsskandal in Nahost* (Gottingen: Steidl Verlag, 1991).

72. United Nations, *Transparency in International Arms Transfers,* UN Disarmament Topical Papers, no. 3 (New York: United Nations, 1990).

73. "Bulgarians to Share Data on Arms Sent to Terrorists," *New York Times*, 2 August 1990; David Godhart and Andrew Fisher, "Germany's Trade Surplus Down by 20 Per Cent," *Financial Times*, 15 February 1991, 8; "French to List Export Details," *Jane's Defence Weekly*, 11 May 1991, 775; "Belousov Details 'Diminished' Military Exports," *Tass*. English translation in *FBIS-SOV-91-006* (Washington: Foreign Broadcast Information Service 1991), 45–46.

74. Naoki Usui, "Kaifu Calls on UN to Monitor Conventional Arms," *Defense News*, 3 June 1991; Robert Mauthner, "Leaders Call for Register on International Arms Sales," *Financial Times*, 17 July 1991, 4.

75. "Unify Arms Control," *Defense News*, 22 April 1991, 18.

76. Giovanni de Briganti, "EC Ponders Single Policy to Regulate Arms Sales," *Defense News*, 1 April 1991, 1.

77. Giovanni de Briganti, "European Governments Take Steps to Tighten Military Export Controls," *Defense News*, 1 April 1991, 20; Allessandro Politi, "Italians Seek Global Forum on Arms Sales," *Defense News*, 11 March 1991, 4; Giovanni de Briganti, "France to Urge Export Policy Coordination," *Defense News*, 8 April 1991, 1; Ernie Regehr, "Canada Prods United States on Arms Sales," *Arms Control Today* (June 1991): 14–16.

78. Alan Cowell, "Egypt Proposes Regional Arms Control Plan," *New York Times*, 5 July 1991, A6.

79. Giovanni di Briganti, "EC Ponders Single Policy to Regulate Arms Sales," *Defense News*, 1 April 1991, 6; Jac Lewis, "EC Export Control Scheme Planned," *Jane's Defence Weekly*, 8 June 1991, 987.

80. *Middle East Arms Control Initiative* (Washington: White House, Office of the Press Secretary. 29 May 1991).

81. For a complete list, see "Congress's Actions on Arms Transfers: From Limits to Loans," *Arms Control Today* (June 1991): 18–19.

82. For a description of the focused comparison method that fosters such theory development, see Alexander George, "Case Studies and Theory Development: The Method of Structured, Focused Comparison," in *Diplomatic History: New Approaches,* ed. Paul Gorden Lauren (New York: Free Press, 1979), 43–68.

# Bibliography

The items in this bibliography were selected first of all because of their availability in most libraries, although there are some exceptions. Secondly, although U.S. government documents are a major source of primary data and information, this bibliography for several reasons does not attempt to make a thorough survey of these sources. Modern search capabilities in libraries now make it possible to locate well-indexed government documents; also, the bibliographies contained in several of the major books on U.S. arms transfers list the major studies that have been conducted. The readers of this volume are reminded that congressional hearings, the *Congressional Record*, and the General Accounting Office produce and publish extremely useful analyses. Finally, the extensive literature that exists on individual suppliers and recipients has not been included. Rather, the focus is on those sources that can be used to provide further background and evidence for the systemic analysis conducted in this book.

## Bibliographies of the International Arms Trade

Ball, Nicole. "The Arms Trade: A Selected Bibliography." In *Arms Transfers in the Modern World*. ed. Stephanie G. Neuman and Robert E. Harkavy, 323–61. New York: Praeger Publishers, 1979.

*Conventional Arms Transfers: A Bibliography,* Washington: Arms Control Association, 1979.

Catrina, Christian. "References." In *Arms Transfers and Dependence*, 395–409. New York: Taylor and Francis, 1988.

Gillingham, Arthur. *Arms Traffic: An Introduction and Bibliography.* Political Issues Series, vol. 4, no. 2. Los Angeles: Bibliographic Reference Service, Center for the Study of Armament and Disarmament, California State University, 1976.

Hammond, Paul Y., David J. Louscher, Michael D. Salomone, and Norman A. Graham. "Selected Bibliography." In *The Reluctant Supplier: U.S. Decisionmaking for Arms Sales.* 277–96. Cambridge: Oelgeschlager, Gunn & Hain, Publishers, 1983.

Harkavy, Robert E. "Bibliography." In *The Arms Trade and International Systems,* 277–88. Cambridge: Ballinger Publishing Company, 1975.

Klare, Michael T. "Selected Bibliography." In *American Arms Supermarket,* 297–302. Austin: University of Texas Press, 1984.

Lumpe, Lora. "Selected Sources on Arms Transfers and Restraints." *Arms Control Today* (June 1991): 37–39.
"Selective Bibliography." In *Arms Transfers to the Third World: 1971–85*, ed. Michael Brzoska and Thomas Ohlson, 370–75. New York: Oxford University Press, 1987.

## Basic Books

The following list of books are those that any researcher in this field should consider part of his or her library. They contain information, data, and analysis that remain useful despite the passage of time. Although some are out of print, they are valuable enough to obtain and copy.

Adams, James. *Engines of War: Merchants of Death and the New Arms Race*. New York: Atlantic Monthly Press, 1990.
Anthony, Ian. *The Naval Arms Trade*. Oxford: Oxford University Press, 1990.
Ball, Nicole. *Briefing Book on Conventional Arms Transfers*. Boston: Council for a Livable World Education Fund, 1991.
Brzoska, Michael, and Thomas Ohlson, eds. *Arms Transfers to the Third World: 1971–85*. New York: Oxford University Press, 1987.
Brzoska, Michael, and Thomas Ohlson, eds. *Arms Production in the Third World*. New York: Taylor and Francis, 1986.
Burns, Richard Dean, ed. *The International Trade in Armaments Prior to World War II*. New York: Garland Publishing, 1972.
Cannizzo, Cindy, ed. *The Gun Merchants: Politics and Policies of the Major Arms Suppliers*. Elmsford, New York: Pergamon Press, 1980.
Catrina, Christian. *Arms Transfers and Dependence*. New York: Taylor and Francis, 1988.
Collier, Basil. *Arms and the Men: The Arms Trade and Governments*. London: Hamish Hamilton, 1980.
Copper, John F., and Daniel S. Papp, eds. *Communist Nations' Military Assistance*. Boulder, Colo.: Westview Press, 1983.
Frank, Lewis A. *The Arms Trade in International Relations*. New York: Praeger Publishers, 1969.
Graves, Ernest, and Steven A. Hildreth, eds. *U.S. Security Assistance: The Political Process*, Lexington, Mass.: Lexington Books, 1985.
Hammond, Paul Y., David J. Louscher, Michael D. Salomone, and Norman A. Graham. *The Reluctant Supplier: U.S. Decisionmaking for Arms Sales*. Cambridge, Mass.: Oelgeschlager, Gunn & Hain, Publishers, 1983.
Harkavy, Robert E. *The Arms Trade and International Systems*. Cambridge: Ballinger Publishing Company, 1975.
Howe, Russell Warren. *Weapons: The International Game of Arms, Money and Diplomacy*. Garden City, N.Y.: Doubleday & Company, 1980.
Jones, Rodney W., and Steven A. Hildreth. *Modern Weapons and Third World Powers*. Boulder, Colo.: Westview Press, 1984.
Katz, James Everett, ed. *Arms Production in Developing Countries*. Lexington, Mass.: Lexington Books, 1984.
Klare, Michael T. *American Arms Supermarket*. Austin: University of Texas Press, 1984.

Kolodziej, Edward A. *Making and Marketing Arms: The French Experience and Its Implications for the International System*. Princeton: Princeton University Press, 1987.
Krause, Keith. *Weapons between States: The Arms Trade in International Relations*. Cambridge: Cambridge University Press, 1992.
Leiss, Amelia C., et al. *Arms Transfers to Less Developed Countries*. Arms Control and Local Conflict, vol. 3. Cambridge: Center for International Studies, MIT, 1970.
Louscher, David J., and Michael D. Salomone, eds. *Marketing Security Assistance: New Perspectives on Arms Sales*. Lexington, Mass.: Lexington Books, 1987.
Neuman, Stephanie G. *Military Assistance in Recent Wars: The Dominance of the Superpowers*. New York: Praeger Publishers, 1986.
Neuman, Stephanie G., and Robert E. Harkavy, eds. *Arms Transfers in the Modern World*. New York: Praeger Publishers, 1979.
Nolan, Janne. *Trappings of Power: Ballistic Missiles in the Third World*. Washington: Brookings Institution, 1991.
Ohlson, Thomas, ed. *Arms Transfer Limitations and Third World Security*. New York: Oxford University Press, 1988.
Pierre, Andrew J., ed. *Arms Transfers and American Foreign Policy*. New York: New York University Press, 1979.
———. *The Global Politics of Arms Sales*. Princeton: Princeton University Press, 1982.
Ra'anan, Uri, Robert L. Pfaltzgraff, Jr., and Geoffrey Kemp, eds. *Arms Transfers to the Third World: The Military Buildup in Less Industrial Countries*. Boulder, Colo.: Westview Press, 1978.
Sampson, Anthony. *The Arms Bazaar*. New York: Viking Press, 1977.
Stockholm International Peace Research Institute. *The Arms Trade with the Third World*. New York: Humanities Press, 1971.
Stanley, John, and Maurice Pearton. *The International Trade in Arms*. New York: Praeger Publishers, 1972.
Sutton, John L., and Geoffrey Kemp. *Arms to Developing Countries: 1945-1965*. Adelphi Papers, no. 28. London: Institute for Strategic Studies, 1966.
Thayer, George. *The War Business: The International Trade in Armaments*. London: Paladin, 1970.
United Nations. *Transparency in International Arms Transfers*. UN Disarmament Topical Papers, no. 3. New York: United Nations, 1990.
U.S. Congress, Office of Technology Assessment. *Global Arms Trade*. OTA-ISC-460. Washington: GPO, 1991.

## Arms Trade and the International System

Albrecht, Ulrich, D. Ernest, Peter Lock and Herbert Wulf, "Militarization, Arms Transfer and Arms Production in Peripheral Countries." *Journal of Peace Research* 12, no. 3 (1975): 195-212.
Anthony, Ian. "The Global Arms Trade." *Arms Control Today*, 21, no. 5 (June 1991): 3-8.
Anthony, Ian, Agnes Courades Allebeck, Gerd Hagmeyer-Gaverus, Paolo Miggiano, and Herbert Wulf. "The Trade in Major Conventional Weapons." In *SIPRI Yearbook*

*1991: World Armaments and Disarmament,* 197-231. Oxford: Oxford University Press, 1991.

Brzoska, Michael. "The Impact of Arms Production in the Third World." *Armed Forces and Society,* 15, no. 4 (Summer 1989): 507-30.

Cordesman, Anthony H. "The Soviet Arms Trade: Patterns for the 1980s (Part One)." *Armed Forces Journal* (June 1983): 96-105.

———. "U.S. and Soviet Competition in Arms Exports and Military Assistance." *Armed Forces Journal International* (August 1981): 65-72.

Evans, Carol. "Reappraising Third-World Arms Production." *Survival* (March/April 1986): 99-116.

Ferrell, Robert H. "The Merchants of Death: Then and Now." *International Affairs* 26, no. 1 (1972): 29-39.

Haftendorn, Helga. "The Proliferation of Conventional Arms." *The Diffusion of Power: Proliferation of Force.* Adelphi Papers, no. 133, 33-41. London: International Institute for Strategic Studies, 1977.

Hagelin, Bjorn. "Multinational Weapon Projects and the International Arms Trade." In *SIPRI Yearbook 1984,* 151-63. London: Taylor and Francis, 1985.

Hoagland, John H. "Arms in the Developing World." *Orbis* 12 (Spring 1968): 167-84.

———. "Arms in the Third World." *Orbis* 14 (Summer 1970): 500-504.

Hoagland, John H., Jr., and John B. Teeple. "Regional Stability and Weapons Transfer: The Middle Eastern Case." *Orbis* 9 (Fall 1965): 714-28.

Karp, Aaron. "Ballistic Missiles in the Third World." *International Security* (Winter 1984-85): 166-95.

———. "The Trade in Conventional Weapons." In *SIPRI Yearbook 1988: World Armaments and Disarmament,* 175-263. New York: Oxford University Press, 1988.

Kemp, Geoffrey. "The New Strategic Map." *Survival* 19, no. 2 (March/April 1977): 50-57.

Klare, Michael. "Deadly Convergence: the Perils of the Arms Trade." *World Policy Journal* (Winter 1988-89): 141-68.

———. "The State of the Trade: Global Arms Transfer Patterns in the 1980s." *Journal of International Affairs* 40, no. 1 (Summer 1986): 1-22.

———. "The Unnoticed Arms Trade: Exports of Conventional Arms Making Technology." *International Security* 8, no. 2 (Fall 1983): 69-88.

Kolodziej, Edward A., and Frederick S. Pearson. *The Political Economy of Making and Marketing Arms: A Test for the Systemic Imperatives of Order and Welfare.* Unpublished paper, University of Illinois and University of Missouri, St. Louis, 1989.

Krause, Keith. "The Political Economy of the International Arms Transfer System: The Diffusion of Military Technique via Arms Transfers." *International Journal* 45, no. 3 (Summer 1990): 687-722.

Landgren-Backstrom, Signe. "Global Arms Trade: Scope, Impact, Restraining Action." *Bulletin of Peace Proposals* 13, no. 3 (1982): 201-10.

McCain, John. "Arms Sales and the Third World." *Washington Quarterly* (Spring 1991): 79-89.

Miksche, F. O. "The Arms Race in the Third World." *Orbis* 12, no. 1 (Spring 1968): 161-84.

Miller, Morton. "Conventional Arms Trade in the Developing World, 1976-86: Reflections on a Decade." In *World Military Expenditures and Arms Transfers*

*1987,* U.S. Arms Control and Disarmament Agency, 19-24. Washington: GPO, 1988.

Neuman, Stephanie G. "Arms, Aid and the Superpowers." *Foreign Affairs* 66, no. 5 (Summer 1988): 1044-66.

———. "The Arms Market: Who's on Top?" *Orbis* 33 (Fall 1989): 509-29.

———. "The Arms Trade in Recent Wars: The Role of the Superpowers." *Journal of International Affairs* 40, no. 1 (Summer 1986): 77-100.

———. "Coproduction, Barter and Countertrade: Offsets in the International Arms Market." *Orbis* 29, no. 1 (Spring 1985): 183-214.

———. "International Stratification and Third World Military Industries." *International Organization* 38, no. 1 (Winter 1984): 167-97.

———. "Third World Arms Production and the Global Arms Transfer System." In *Arms Production in Developing Countries,* ed. James Everett Katz, 15-38. Lexington, Mass.: Lexington Books, 1984.

Nolan, Janne. "The Global Arms Market after the Cold War." *Washington Quarterly* 14, no. 3 (1991): 386-423.

Oberg, Jan. "Arms Trade with the Third World as an Aspect of Imperialism." *Journal of Peace Research* 12, no. 3 (1975): 213-34.

Ohlson, Thomas, and Elizabeth Skons. "The Trade in Major Conventional Weapons." In *SIPRI Yearbook 1987,* Stockholm International Peace Research Institute, 181-209. New York: Oxford University Press, 1987.

Peleg, Ilan. "Arms Supply to Third World Countries: Models and Explanations." *Journal of Modern African Studies* 15, no. 1 (1977): 91-103.

———. "Military Production in Third World Countries: A Political Study." *Threats, Weapons, and Foreign Policy,* ed. Pat McGowan and Charles W. Kegley, 210-30. Beverly Hills, Calif.: Sage Publications, 1980.

Pierre, Andrew J. "Arms Sales: The New Diplomacy." *Foreign Affairs* (Winter 1981-82): 266-86.

Rosh, Robert M. "Third World Arms Production and the Evolving Interstate System." *Journal of Conflict Resolution* 34, no. 1 (March 1990): 57-73.

Smith, Ron, Anthony Humm, and Jacques Fontanel. "The Economics of Exporting Arms." *Journal of Peace Research* 2, no. 3 (1985): 239-47.

Tullberg, Rita. "Military-Related Debt in Non-Oil Developing Countries, 1972-82." In *SIPRI Yearbook 1985: World Armaments and Disarmament,* 445-458. London: Taylor and Francis, 1985.

## Measurement, Data, and Methodology

Brzoska, Michael. "Arms Transfer Data Sources." *Journal of Conflict Resolution* 26, no. 1 (March 1982): 77-108.

Fei, Edward T. "Understanding Arms Transfers and Military Expenditures: Data Problems." In *Arms Transfers in the Modern World,* ed. Stephanie G. Neuman and Robert E. Harkavy, 37-48. New York: Praeger Publishers, 1979.

Gail, Bridget. "'The Fine Old Game of Killing,' Comparing U.S. and Soviet Arms Sales." *Armed Forces Journal* (September 1978): 16.

Kolodziej, Edward A. "Measuring French Arms Transfers: A Problem of Sources and

Some Sources of Problems with ACDA Data." *Journal of Conflict Resolution* 23, no. 2 (June 1979): 195–227.

Laurance, Edward J., and Joyce A. Mullen. "Assessing and Analyzing International Arms Trade Data." In *Marketing Security Assistance,* ed. David J. Louscher and Michael D. Salomone, 79–98. Lexington, Mass.: Lexington Books, 1987.

Laurance, Edward J., and Ronald G. Sherwin. "The Measurement of Weapons-System Balances: Building upon the Perceptions of Experts." In *International Perceptions of the Superpower Military Balance,* ed. Donald C. Daniel, 40–53. New York: Praeger Publishers, 1983.

———. "Understanding Arms Transfers through Data Analysis." In *Arms Transfers to the Third World: The Military Buildup in Less Industrial Countries,* ed. Uri Ra'anan, et al., 87–106. Boulder, Colo.: Westview Press, 1978.

Leiss, Amelia C. "International Transfers of Armaments: Can Social Scientists Deal with Qualitative Issues?" In *Arms Transfers to the Third World: The Military Buildup in Less Industrial Countries,* ed. Uri Ra-anan, et al., 107–20. Boulder, Colo.: Westview Press, 1978.

Louscher, David J., and Michael D. Salomone. "The Imperative for a New Look at Arms Sales." In *Marketing Security Assistance,* ed. David J. Louscher and Michael D. Salomone, 13–40. Lexington, Mass.: Lexington Books, 1987.

Mayer, Laurel A. "U.S. Arms Transfers: Data Sources and Dilemmas." *International Studies Notes* 7, no. 2 (Summer 1980): 1–7.

Sherwin, Ronald, and Edward J. Laurance. "Arms Transfers and Military Capability: Measuring and Evaluating Conventional Arms Transfers." *International Studies Quarterly* 23, no. 1 (September 1979): 360–89.

Wilcox, Richard H. "Twixt Cup and Lip: Some Problems in Applying Arms Controls." In *Arms Transfers in the Modern World,* ed. Stephanie G. Neuman and Robert E. Harkavy, 27–36. New York: Praeger Publishers, 1979.

# Norms and Rules of the International Arms Trade System

## Arms Transfers and Conflict

Deger, Saadet. "Regional Conflict and Recent Trends in the International Arms Trade." In *Arms and Defence in Southeast Asia,* ed. Chandran Jeshurun, 152–80. Singapore: Institute of Southeast Asian Studies, 1989.

Dunn, Lewis and James Tomashoff. *New Technologies and the Changing Dimensions of Third World Military Conflict.* CNSN Paper vol. 2, No. 1. McLean, Va.: Science Applications International Corporation, Center for National Security Negotiations, 1990.

Harkavy, Robert E., and Stephanie G. Neuman, eds. *The Lessons of Recent Wars in the Third World: Approaches and Case Studies, Vol. 1.* Lexington, Mass.: Lexington Books, 1985.

Hoagland, John H., Jr., and John B. Teeple. "Regional Stability and Weapons Transfer: The Middle Eastern Case." *Orbis* 9 (Fall 1965): 714–728.

Hudson, C. I. "The Impacts of Precision Guided Munitions on Arms Transfers and

International Stability." In *Arms Transfers in the Modern World*, ed. Stephanie G. Neuman and Robert E. Harkavy, 77-88. New York: Praeger Publishers, 1979.
Krause, Keith. "Conflict Management, Arms Transfers, and the Arab-Israeli Conflict." In *Conflict Management in the Middle East,* ed. Gabriel Ben-Dor and David B. Dewitt, 209-38. Lexington, Mass.: Lexington Books, 1987.
Neuman, Stephanie G. "Arms, Aid and the Superpowers." *Foreign Affairs* 66, no. 5 (Summer 1988): 1044-66.
———. "Arms and Superpower Influence: Lessons From Recent Wars." *Orbis* 31 (Winter 1987): 711-29.
———. "The Arms Trade in Recent Wars: The Role of the Superpowers." *Journal of International Affairs* 40, no. 1 (Summer 1986): 77-100.
———. *Military Assistance in Recent Wars: The Dominance of the Superpowers.* New York: Praeger Publishers, 1986.
———. "Summary of Lessons." *The Lessons of Recent Wars in the Third World: Approaches and Case Studies, Vol. 1,* ed. Robert E. Harkavy and Stephanie G. Neuman, 281-92. Lexington, Mass.: Lexington Books, 1985.
Porter, Bruce D. *The USSR in Third World Conflicts: Soviet Arms and Diplomacy in Local Wars 1945-1980.* Cambridge: Cambridge University Press, 1984.
Slaughter, Ronald. "The Politics and Nature of the Conventional Arms Transfer Process during a Military Engagement: The Falklands-Malvinas Case." *Arms Control* 4 (May 1983): 16-30.

## Arms Transfers, Foreign Policy, Diplomacy, and Influence

Arlinghaus, Bruce E. "Linkage and Leverage in African Arms Transfers." In *Arms for Africa,* ed. Bruce E. Arlinghaus, 3-18. Lexington, Mass.: Lexington Books, 1983.
Becker, Abraham S. *Arms Transfers, Great Power Intervention and Settlement of the Arab-Israeli Conflict.* Santa Monica: The Rand Corporation, 1977.
Cahn, Anne. "United States Arms to the Middle East 1967-76: A Critical Examination." In *Great Power Intervention in the Middle East.* ed. Milton Leitenberg and Gabriel Sheffer 101-33 New York: Pergamon Press, 1979.
Harkavy, Robert E. "The New Geopolitics: Arms Transfers and the Major Powers' Competition for Overseas Bases." In *Arms Transfers in the Modern World,* ed. Stephanie G. Neuman and Robert E. Harkavy, 131-154. New York: Praeger Publishers, 1979.
Karsh, Efraim. "Influence through Arms Supplies: The Soviet Experience in the Middle East." *Conflict Quarterly* (Winter 1986): 45-57.
Krause, Keith. "Conflict Management, Arms Transfers, and the Arab-Israeli Conflict." In *Conflict Management in the Middle East,* ed. Gabriel Ben-Dor and David B. Dewitt, 209-38. Lexington, Mass.: Lexington Books, 1987.
Legvold, Robert. "Soviet and Chinese Influence in Black Africa." In *Soviet and Chinese Influence in the Third World,* ed. Alvin Z. Rubenstein, 154-75. New York: Praeger Publishers, 1975.
Lewis, William H. "Political Influence: The Diminished Capacity." In *Arms Transfers in the Modern World,* ed. Stephanie G. Neuman and Robert E. Harkavy, 184-99. New York, New York: Praeger Publishers, 1979.
Nachmias, Nitza. *Transfer of Arms, Leverage and Peace in the Middle East.* Westport, Conn.: Greenwood Press, 1988.

Quandt, William B. "Influence through Arms Supply: The U.S. Experience in the Middle East." In *Arms Transfers to the Third World: The Military Buildup in Less Industrial Countries,* ed. Uri Ra'anan, et al., 121–30. Boulder, Colo.: Westview Press, 1978.

Ra'anan, Uri. "Soviet Arms Transfers and the Problem of Political Leverage." In *Arms Transfers to the Third World: The Military Buildup in Less Industrial Countries,* ed. Uri Ra'anan, et al., 131–58. Boulder, Colo.: Westview Press, 1978.

Roeder, Philip G. "The Ties That Bind: Aid, Trade, and Political Compliance in Soviet–Third World Relations." *International Studies Quarterly* 29 (1985): 191–216.

Rubenstein, Alvin Z. "Assessing Influence as a Problem in Foreign Policy Analysis." In *Soviet and Chinese Influence in the Third World,* ed. Alvin Z. Rubenstein, 1–21. New York: Praeger Publishers, 1975.

———. *Red Star on the Nile: The Soviet-Egyptian Influence Relationship since the June War.* Princeton: Princeton University Press, 1977.

Wheelock, Thomas R. "Arms for Israel: The Limit of Leverage." *International Security* 3, no. 2 (Fall 1978): 123–37.

## Arms Transfers and Strategic Resources

Beker, Avi. "The Arms-Oil Connection: Fueling the Arms Race." *Armed Forces and Society* 3, no. 3 (Spring 1982): 419–42.

Gawad, Atef A. "Moscow's Arms-for-Oil Diplomacy." *Foreign Policy* (Summer 1986): 147–68.

Laurance, Edward J. "An Assessment of the Arms-for-Oil Strategy." In *Energy and National Security,* ed. Donald J. Goldstein, 59–92. Washington: National Defense University Press, 1981.

Snider, Lewis W. "Arms Exports for Oil Imports? The Test of a Nonlinear Model." *Journal of Conflict Resolution* 28, no. 4 (December 1984): 680–97.

## Economic Effects of International Arms Transfers

Arlinghaus, Bruce E. "Social versus Military Development: Positive and Normative Dimensions." In *Arms Production in Developing Countries,* ed. James Everett Katz, 39–50. Lexington, Mass.: Lexington Books, 1984.

Chan, Steve. "The Impact of Defense Spending on Economic Performance: A Survey of Evidence and Problems." *Orbis* 29, no. 2 (Summer 1985): 403–34.

Lock, Peter, and Herbert Wulf. "Consequences of the Transfer of Military-oriented Technology on the Development Process." *Bulletin of Peace Proposals* 8, no. 2 (1977): 127–36.

Looney, Robert E. *Third-World Military Expenditure and Arms Production.* London: Macmillan Press, 1988.

Neuman, Stephanie G. "Arms Transfers and Economic Development: Some Research and Policy Issues." In *Arms Transfers in the Modern World,* ed. Stephanie G. Neuman and Robert E. Harkavy, 219–45. New York: Praeger Publishers, 1979.

Tullberg, Rita. "Military-Related Debt in Non-Oil Developing Countries, 1972–82." In *SIPRI Yearbook 1985: World Armaments and Disarmament,* 445–58. London: Taylor and Francis, 1985.

## Arms Control and the Arms Trade

Alagappa, Muthiah, and Noordin Sopiee. "Problems and Prospects for Arms Control in South-East Asia." In *Arms Transfer Limitations and Third World Security*, ed. Thomas Ohlson, 186-97. Oxford: Oxford University Press, 1988.
Anthony, Ian, ed. *Arms Export Regulations*. Oxford: Oxford University Press, 1991.
Ball, Nicole. "Third World Arms Control: A Third World Responsibility." In *Arms Transfer Limitations and Third World Security*, ed. Thomas Ohlson, 45-57. Oxford: Oxford University Press, 1988.
Betts, Richard. "The Tragicomedy of Arms Trade Control." *International Security* (Summer 1980): 80-110.
Blechman, Barry M., Janne Nolan, and Alan Platt. "Pushing Arms." *Foreign Policy* 46 (Spring 1982): 138-54.
Bloomfield, Lincoln P., and Amelia Leiss. "Arms Transfer and Arms Control." *Proceedings of the American Academy of Political Science* 29 (March 1969): 37-54.
———. *Controlling Small Wars: A Strategy for the 1970s*. New York: Alfred A. Knopf, 1969.
Cahn, Anne Hessing. "Arms Transfer Constraints." In *Arms Transfers to the Third World: The Military Buildup in Less Industrial Countries*, ed. Uri Ra'anan, et al., 327-44. Boulder, Colo.: Westview Press, 1978.
Cahn, Anne Hessing, Joseph J. Kruzel, Peter M. Dawkins and Jacques Huntzinger. et al., eds. *Controlling Future Arms Trade*. New York: McGraw-Hill Book Company, 1977.
Cannizzo, Cindy "Prospects for the Control of Conventional Arms Transfers." In *The Gun Merchants*, ed. Cindy Cannizzo, 187-96. Elmsford, N.Y.: Pergamon Press, 1980.
Child, Jack. "Interstate Conflict, Conflict Resolution and Arms Transfers." *Latin American Research Review* (1985): 172-181.
Fontanel, Jacques, and Jean-Francois Guilhaudis. "Arms Transfer Control and Proposals to Link Disarmament to Development." In *Arms Transfer Limitations and Third World Security*, ed. Thomas Ohlson, 215-26. Oxford: Oxford University Press, 1988.
Franko, Lawrence. "Restraining Arms Exports to the Third World: Will Europe Agree?" *Survival* 21, no. 1 (January/February 1979): 14-25.
Gray, Colin. "Traffic Control for the Arms Trade." *Foreign Policy* (Spring 1972): 153-69.
Hagelin, Bjorn. "Arms Transfer Limitations: The Case of Sweden." *Arms Transfer Limitations and Third World Security*, ed. Thomas Ohlson, 157-72. Oxford: Oxford University Press, 1988.
Hettne, Bjorn. "Third World Arms Control and World System Conflicts." In *Arms Transfer Limitations and Third World Security*, ed. Thomas Ohlson, 17-32. Oxford: Oxford University Press, 1988.
Hunter, Robert E. "Arms Control in the Persian Gulf." *Arms Transfers and American Foreign Policy*, ed. Andrew J. Pierre, 98-120. New York: New York University Press, 1979.
Husbands, Jo L., and Anne Cahn. "The Conventional Arms Transfers Talks: An Experiment in Mutual Arms Trade Restraint." In *Arms Transfer Limitations and Third World Security*, ed. Thomas Ohlson, 110-25. Oxford: Oxford University Press, 1988.

Jabbar, Paul. *Not by War Alone: Security and Arms Control in the Middle East.* Berkeley: University of California Press, 1981.

Kemp, Geoffrey. "The Military Build-Up: Arms Control or Arms Trade?" *The Middle East and the International System. Part I: The Impact of the 1973 War.* Adelphi Papers, no. 114. London: Institute for Strategic Studies, 1975.

Klare, Michael T. "An Arms Control Agenda in the Third World." *Arms Control Today* (April 1990): 8–12.

———. "Deadly Convergence: The Perils of the Arms Trade." *World Policy Journal* (Winter 1988–89): 141–168.

———. "Evading the Embargo: Illicit U.S. Arms Transfers to South Africa." *Journal of International Affairs* 35, no. 1 (Spring/Summer 1981): 15–28.

———. "Gaining Control: Building a Comprehensive Arms Restraint System." *Arms Control Today* 21, no. 5 (June 1991): 9–13.

Krause, Joachim. "Soviet Arms Transfer Restraint." In *Arms Transfer Limitations and Third World Security,* ed. Thomas Ohlson, 95–109. Oxford: Oxford University Press, 1988.

Krause, Keith. "Constructing Regional Security Regimes and the Control of Arms Transfers." *International Journal* 45 (Spring 1990): 386–423.

Muni, S. D. "Third World Arms Control: Role of the Non-aligned Movement." In *Arms Transfer Limitations and Third World Security,* ed. Thomas Ohlson, 198–212. Oxford: Oxford University Press, 1988.

Nolan, Janne E. "The U.S.-Soviet Conventional Arms Transfer Negotiations." In *U.S.-Soviet Security Cooperation,* ed. Alexander George, Philip J. Farley and Alexander Dallin, 510–23. New York: Oxford University Press, 1988.

Nolan, Janne. "The Global Arms Market after the Cold War." *Washington Quarterly* 14, no. 3 (1991): 386–423.

Ohlson, Thomas, ed. *Arms Transfer Limitations and Third World Security.* Oxford: Oxford University Press, 1988.

Pearson, Frederick S. "Problems and Prospects of Arms Transfer Limitations among Second-Tier Suppliers: The Cases of France, the United Kingdom and the Federal Republic of Germany." In *Arms Transfer Limitations and Third World Security,* ed. Thomas Ohlson, Oxford: Oxford University Press, 1988. 126–56.

———. "The Question of Control in British Defence Sales Policy." *International Affairs* 59, no. 2 (Spring 1983): 211–38.

Pierre, Andrew J. "Multilateral Restraints on Arms Transfers." In *Arms Transfers and American Foreign Policy,* ed. Andrew J. Pierre, 228–322. New York: New York University Press, 1979.

"The Proposal for an Arms Trade Register." In *The International Trade in Arms: Problems and Prospects.* Ottawa: Canadian Institute for International Peace and Security, 1988.

Salomone, Michael D., and David J. Louscher. "Lessons of the Carter Approach to Restraining Arms Transfers." *Survival* (September/October 1981): 200–208.

Simpson, John. "The Nuclear Non-proliferation Regime as a Model for Conventional Armament Restraint." In *Arms Transfer Limitations and Third World Security,* ed. Thomas Ohlson, 227–41. Oxford: Oxford University Press, 1988.

Sloutsky, Naoum. *The League of Nations and the Control of Trade in Arms.* Report No. C/74-8. Cambridge: Center for International Studies, MIT, 1974.

Smith, Chris. "Third World Arms Control, Military Technology and Alternative

Security." In *Arms Transfer Limitations and Third World Security,* ed. Thomas Ohlson, 58–72. Oxford: Oxford University Press, 1988.
Subrahmanyam, K. "Third World Arms Control in a Hegemonistic World." In *Arms Transfer Limitations and Third World Security,* ed. Thomas Ohlson, 33–44. Oxford: Oxford University Press, 1988.
Taylor, Trevor. "The Evaluation of Arms Transfer Control Proposals." In *The Gun Merchants,* ed. Cindy Cannizzo, 167–86. Elmsford, N.Y.: Pergamon Press, 1980.
*Transparency in International Arms Transfers.* UN Disarmament Topical Papers, no. 3. New York: United Nations, 1990.
Varas, Augusto. "Regional Arms Control in the South American Context." In *Arms Transfer Limitations and Third World Security,* ed. Thomas Ohlson, 175–185. Oxford: Oxford University Press, 1988.
Wulf, Herbert. "Arms Transfer Control: The Feasibility and the Obstacles." In *Defence, Security and Development,* ed. Saadet Deger and Robert West, London: Frances Pinter Publishers, 1987. 191–214.

## Ballistic Missile Proliferation

Aspen Strategy Group. *New Threats: Responding to the Proliferation of Nuclear, Chemical, and Delivery Capabilities in the Third World.* Lanham, Md.: University Press of America, 1990.
*Assessing Ballistic Missile Proliferation and Its Control.* Stanford: Center for International Security and Arms Control, Stanford University, 1991.
Carus, Seth. *Ballistic Missiles in the Third World: Threat and Response.* Report. Washington: Center for International and Strategic Studies, 1990.
Hollinger, Frederick J. "The Missile Technology Control Regime: A Major New Arms Control Achievement." In *World Military Expenditures and Arms Transfers 1987,* U.S. Arms Control and Disarmament Agency, 25–28. Washington: GPO, March 1988.
Karp, Aaron. "Ballistic Missiles in the Third World." *International Security* (Winter 1984–85): 166–95.
———. "Ballistic Missile Proliferation in the Third World." In *SIPRI Yearbook 1989: World Armaments and Disarmament,* 287–318. Oxford: Oxford University Press, 1989.
———. "Ballistic Missile Proliferation." *SIPRI Yearbook 1990: World Armaments and Disarmament,* 368–91. Oxford: Oxford University Press, 1990.
———. "Ballistic Missile Proliferation." *SIPRI Yearbook 1991: World Armaments and Disarmament,* 317–44. Oxford: Oxford University Press, 1991.
———. "The Frantic Third World Quest for Ballistic Missiles." *Bulletin of Atomic Scientists* (June 1988): 14–19.
Mahnken, Thomas G., and Timothy Hoyt. "Missile Proliferation and American Interests." *SAIS Review* (Winter/Spring 1990): 101–112.
Mahnken, Thomas G., and Timothy D. Hoyt. "The Spread of Missile Technology to the Third World." *Comparative Strategy* 9, no. 3 (1990): 245–63.
McNaugher, Thomas L. "Ballistic Missiles and Chemical Weapons: The Legacy of the Iran-Iraq War." *International Security* 15, no. 2 (Fall 1990): 5–34.

Milholin, Gary. *Licensing Mass Destruction: U.S. Exports to Iraq, 1985–1990.* Washington: Wisconsin Project on Nuclear Arms Control, 1991.

Milholin, Gary, and Gerard White. *Bombs from Beijing: A Report on China's Nuclear and Missile Exports.* Washington: Wisconsin Project on Nuclear Arms Control, 1991.

"Missile Proliferation in the Third World." In *Strategic Survey 1988–1989*, 14–25. London: International Institute for Strategic Studies, 1989.

Navias, Martin S. "Ballistic Missile Proliferation in the Middle East." *Survival* 31, no. 3 (May/June 1989): 225–40.

Navias, Martin. *Ballistic Missile Proliferation in the Third World.* Adelphi Papers, no. 252. London: International Institute for Strategic Studies, 1990.

Nolan, Janne. "Ballistic Missiles in the Third World—The Limits of Nonproliferation." *Arms Control Today* (November 1989): 9–14.

———. "Missile Mania: Some Rules for the Game." *Bulletin of the Atomic Scientists* 46, no. 4 (May 1990): 27–29.

———. *Trappings of Power: Ballistic Missiles in the Third World.* Washington: Brookings Institution, 1991.

Nolan, Janne E., and Albert D. Wheelon. "Third World Ballistic Missiles." *Scientific American* 263, no. 2 (August 1990): 34–40.

Ottenberg, Michael A. "Operational Implications of Middle East Ballistic Missile Proliferation." *Defense Analysis* 7, no. 1 (1991): 3–19.

Palevitz, Marc S. "Beyond Deterrence: What the U.S. Should Do about Ballistic Missiles in the Third World." *Strategic Review* 18, no. 3 (Summer 1990): 49–58.

Potter, William C., and Adam N. Stulberg. "The Soviet Union and the Spread of Ballistic Missiles." *Survival* 32, no. 6 (November/December 1990): 543–57.

Rubenson, David, and Anna Slomovic. *The Impact of Missile Proliferation on U.S. Power Projection Capabilities.* RAND Report N-2985-A/OSD. Santa Monica, Calif.: RAND Corporation, 1990.

Schmidt, Rachel. *U.S. Export Control Policy and the Missile Technology Control Regime.* RAND Report P-7615-RGS. Santa Monica, Calif.: Rand Graduate School, 1990.

Shuey, Robert D. *Missile Proliferation: Survey of Emerging Missile Forces.* Report. Washington: Congressional Research Service, 1988.

Timmerman, Kenneth R. *The Poison Gas Connection: Western Suppliers of Unconventional Weapons and Technologies to Iraq and Libya.* Los Angeles: Simon Wiesenthal Center, 1990.

Timmerman, Kenneth R. *The Death Lobby: How the West Armed Iraq.* Boston: Houghton Mifflin Company, 1991.

# Index

## -A-

Adams, Gordon, 201–202
Afghanistan-Soviet conflict, 156–157, 190
Aircraft: determining valid and reliable data for fighter, 21–22; military capability and costs of, 38; proliferation of, 143–144
American League for Exports and Security Assistance (ALESA), 188
American League for International Security Assistance, 188
Arens, Moshe, 175
Argentina, as a second-tier supplier, 179
Arms Export Control Act (1976), 6, 7
*Arms Production in the Third World* (SIPRI), 8
*Arms Sales Monitor, The*, 7
*Arms Trade and International Systems, The*, (Harkavy), 12, 49, 58
*Arms Trade with the Third World, The*, (SIPRI), 11, 12
*Arms Transfer Limitations and Third World Security* (SIPRI), 8–9
Arms transfers/trade: defining, 3; events influenced by, 1–3; negative effects of, 5–7; other terms for, 3; political science developments and, 11–12; previous research areas on, 7–10; questions and propositions proposed, 12–15; reasons for, 4–5
Arms transfers/trade, conceptualization and measurement of: basic components of, 16–18; case example of Iraq, 18–20; data problems, 23–30; dollar values, 27–30; importance of reliable and valid data, 20–23; measurement problems, 30, 34–40, 42–44; problems with using specific cases, 40–41

*Arms Transfers to Less Developed Countries* (Bloomfield and Leiss), 11
*Arms Transfers to the Third World: 1971–85* (SIPRI), 8–9, 206
*Arms Transfers to the Third World: The Military Buildup in Less Industrial Countries*, 77–78
Association of Southeast Asian Nations (ASEAN), 197
Atwood, Donald, 199, 200

## -B-

Balance-of-power concept, 53
Ballistic missile proliferation, 129–130, 144–145, 165–166
BCCI bank scandal, 130
Behavior, explaining changes in: arming of Iraq, 182–185; foreign policy influenced by arms transfers, 189–190; illegal arms exports from the U.S., 177–179; response of recipients to, 181–182; second-tier suppliers, 130–132, 137–138, 179–181; Soviet Union as supplier, 172–173; U.S. defense industries, 185–189; U.S. response to offsets, 176–77; U.S.-Soviet behavior as a regime, 54; U.S. as supplier, 173–176
Bloomfield, Lincoln, 11
Boundaries and environment: Holsti framework and, 49–50; 1930–40 period, 58–62; 1946–66 period, 77–81; 1966–80 period, 99–100; 1980–92 period, 125–128; 1992– period, 196–197
Brazil: benefits of systemic knowledge and, 46; as a second-tier supplier, 179
Buchwald, Art, 7
Bush, George, 204

239

240 • Index

-C-

Camp David Accords, 2
Carter administration: research conducted on arms transfers and, 7-8; restraint policy of, 6, 22, 120, 121, 174, 188
Cheney, Dick, 174-175
China: benefits of systemic knowledge and, 45-46; as a second-tier supplier, 180; U.S. as supplier to, 175-176
Cold War: contest for clients during, 2. *See also* Postwar years (1946-66); Postwar years (1966-80); Postwar years (1980-85)
Commission for the Regulation of Arms and Implements of War (1932), 72
Condor II missile, 129-130, 144, 166, 179
Conference for the Control of International Traffic in Arms, Munitions and Implements of War (1925), 7
Conflict, arms exports as a cause of: 1930-40 period, 72-73; 1946-66 period, 94-95; 1966-80 period, 110-114; 1980-92 period, 155-159; 1992- period, 202
Congressional Reference Service (CRS), 26, 30, 44, 136
Construct validity, 37
Continuity and change, system transformation and, 190; historical view, 192-196; verification of, 191-192
Control of arms trade: 1946-66 period, 96-98; 1966-80 period, 120-121; 1980-92 period, 163-167; 1992- period, 203-205
Conventional Armed Forces in Europe Treaty (CFE), 126, 147
Convention for the Control of the Trade in Arms and Ammunition of St. Germain, 72
Coordinating Committee (COCOM), 93-94, 146, 155
Coproduction: 1946-66 period, 94; 1966-80 period, 119; 1992- period, 203

Copying of arms production, 1930-40 period, 70
Countertrade, 43
Counting rules, problems with, 34
Czechoslovakia, as a second-tier supplier, 180-181

-D-

Data: arms trade and historical use of, 23-24; counting rules, 34; importance of reliable and valid, 20-23; reliability, 34-36; secrecy of, 35; sets, 27-30; sources of, 24-27; validity, 36-40
Defense industries, changes in U.S., 185-189
*Defense News,* 26
Defense Production Act (U.S.), 175
Delivery stage: description of, 17-18; Iraq example and, 20
Dollar value: description of, 16-17; Iraq example and, 18; recorded in Congressional Reference Service, 30; recorded in *Foreign Military Sales, Foreign Military Construction Sales and Military Assistance Facts,* 30; recorded in *Military Balance,* 30; recorded in *World Military Expenditures and Arms Transfers,* 27-29; recorded in *Yearbook on Armaments and Disarmament,* 29-30; reliability problems, 34-35

-E-

East/West Engineering Co., Ltd., 173
East-West trade as a regime, 54
Economics, influence of arms trade: 1930-40 period, 59, 61, 74-75; 1946-66 period, 96; 1966-80 period, 100, 118-119; 1980-92 period, 127-128, 161-163; 1992- period, 197, 202-203
Egypt, changes in acquisitions, 181-182
Embargoes, effects of, 2, 74, 113, 121, 166, 174, 179, 184
End-user certificates: creation of, 186; Israel and, 46; 1980-92 period, 131; used by the U.S. and the USSR, 48

Environment. *See* Boundaries and environment
Equipment: dual-use, 155; Iraq example and, 18–19; modernity, 17; number of units transferred, 17; surplus of, 146–147; type and model of, 17
Eurofighter, 132
European Community (EC), 150, 151, 166
European Defense Community (EDC), 192
European Economic Community (EEC), 192
European Fighter Aircraft (EFA), 132
Exhibitions/bazaars, arms, 153
Export-Import Bank, 175
Exports, illegal versus private, 130

-F-

F-16 aircraft, 43, 108, 151
Face validity, 37
Falklands, 156
Federation of American Scientists, 7
FMC Corp., 180
Foreign military sales (FMS), 91, 93
*Foreign Military Sales, Foreign Military Construction Sales and Military Assistance Facts* (DSAA), 26; dollar values recorded in, 30
Foreign policy and diplomacy, arms transfers and political influence, 189–190; 1930–40 period, 73–74; 1946–66 period, 95; 1966–80 period, 115–118; 1980–92 period, 159–161; 1992- period, 202
France, 36, 81, 82, 118, 121, 144, 149, 152
Fulbright, J. William, 187

-G-

G7 countries and arms trade control, 10, 126, 165, 209
General Assembly Resolution 46/36L, 204
General Dynamics, 180
Germany (West), 18, 21–22, 35, 128, 132–133, 149, 161, 166
*Global Arms Trade*, 128, 153
Gray market trade, 154–155

-H-

Harkavy, Robert, 12, 49, 58
Holsti, K. J., 48
Holsti framework: boundaries and environment, 49–50; characteristics of political units, 50–51; modes of interaction, 52–53; regimes, 53–56; structure and stratification, 51; summary of, 56–57
Hussein, Saddam, 183

-I-

Illegal arms trade: measurement problems, 42; reemergence of, 130, 153–154, 161, 166, 198; absence of during interwar years (1930–40), 70; from the U.S., 177–179; versus private, 130
Independent European Program Group (IEPG), 150
Industrial sources of data, 24–25
Intelligence agencies as sources of data, 25
Interaction, modes of: Holsti framework and, 52–53; 1930–40 period, 69–70; 1946–66 period, 91–94; 1966–80 period, 108–110; 1980–92 period, 147–155; 1992- period, 201–202
International Atomic Energy Agency, 55
*International Defense Review*, 130
International Institute for Strategic Studies (IISS), *Military Balance, The*, 24, 26, 30, 39
International Monetary Fund, 128
*International Politics* (Holsti), 48
Interwar years (1930–40): boundaries and environment, 58–62; characteristics of political units, 62–64; conflict, causes of, 72–73; economics and the influence of arms trade, 59, 61, 74–75; foreign policy and diplomacy, influenced by arms trade, 73–74; modes of interaction, 69–70; regime norms and rules, 70, 72–75; structure and stratification, 64–69; summary of, 75–76
Intracolonial trade, 64
Iran-Contra, 178

242 · *Index*

Iran-Iraq war, 127, 137–138, 157–158, 160–161, 171
Iraq: arming of, 182–185; basic dimensions of arms trade and, 18–20; benefits of systemic knowledge and, 46–47
Israel: benefits of systemic knowledge and, 46; as a second-tier supplier, 179; U.S. as supplier to, 175

**-J-**

*Jane's Defence Weekly*, 26, 153
Jane's Publishing Co., 25
*Jane's Weapons Systems*, 146
Japan, 83, 94, 96, 134, 139, 144, 155, 161

**-K-**

Kennedy administration, counter-insurgency during, 97
Kissinger, Henry, 187
Kuhn, Thomas, 12

**-L-**

League of Nations, 5, 23, 61, 64, 185
Leiss, Amelia, 11
Lend-lease agreements, World War II and, 2
Licensing of arms production: 1930–40 period, 69–70, 71; 1946–66 period, 94; 1966–80 period, 110, 119
Lockheed, 189
Logistics and support services: description of, 17; Iraq example and, 19
Low-intensity conflict, 42

**-M-**

Market shares: 1930–40 period, 64–66, 67; 1946–66 period, 86, 87–88; 1966–80 period, 103–104; 1980–92 period, 136–138
Market trade, gray, 154–155
Marshall Plan, 81, 82, 99
Massachusetts Institute of Technology (MIT), Center for International Studies at, 11
"Merchants of death" era, 59, 72–73, 186

"Middle East Arms Control Initiative," 204–205
Military Assistance Advisory Groups (MAAGs), 96
Military Assistance Program (MAP), 82, 91, 92, 96
*Military Balance, The*, (IISS), 24, 26, 30, 39
Military utility: description of, 17; Iraq example and, 19; multiattribute utility and, 39–40; 1930–40 period, 61, 69; 1946–66 period, 90–91; 1966–80 period, 108; 1980–92 period, 139–147
Missile Technology Control Regime (MTCR), 10, 130, 144, 158, 165, 166
Modernity, equipment: description of, 17; Iraq example and, 19; 1930–40 period, 69; 1946–66 period, 81
Multiattribute utility, 39–40
Multinational corporations, 102, 132–33
Munitions-control list, 186

**-N-**

Naval proliferation, 141–143
Near East Arms Coordinating Committee, 97
Nelson Amendment, 187
Nixon Doctrine, 186–187
Non-Proliferation Treaty (1968), 55
North Korea, illegal exports to, 153, 177
Nuclear proliferation as a regime, 54–55
Nunn Amendment, 126
Nye committee hearings. *See* U.S. Senate Munitions Inquiry of 1934–36

**-O-**

Offsets: defining, 43; 1980–92 period, 125, 151–153; 1992- period, 201; problems presented by, 43–44; U.S. response to, 176–77
Off-the-shelf deliveries, 93, 203
OPEC on petroleum dollars, 72, 100, 118, 128

**-P-**

Pakistan, U.S. as supplier to, 175
Panavia, 22, 132
Paris Peace Conference (1919), 72

Payment, mode of: description of, 17, 52; Iraq example and, 20; 1930–40 period, 69; 1946–66 period, 82, 91–93; 1966–80 period, 108–109; 1980–92 period, 147; 1992- period, 201
Persian Gulf War, 3, 6, 142, 143, 156, 162, 164; arming of Iraq, 182–185
Political science developments, research on arms trade and, 11–12
Political units, characteristics of: Holsti framework and, 50–51; 1930–40 period, 62–64; 1946–66 period, 81–84; 1966–80 period, 100–102; 1980–92 period, 128–133; 1992- period, 198
Political visibility, 161
Postwar years (1946–66): boundaries and environment, 77–81; characteristics of political units, 81–84; conflict, causes of, 94–95; control of arms trade, 96–98; economics and the influence of arms trade, 96; foreign policy and diplomacy, influenced by arms trade, 95; modes of interaction, 91–94; regime norms and rules, 94–98; structure and stratification, 85–90; summary of, 98
Postwar years (1966–80): boundaries and environment, 99–100; characteristics of political units, 100–102; conflict, causes of, 110–114; control of arms trade, 120–121; economics and the influence of arms trade, 100, 118–119; foreign policy and diplomacy, influenced by arms trade, 115–118; modes of interaction, 108–110; regime norms and rules, 110–121; structure and stratification, 103–108; summary of, 121–123
Postwar years (1980–92): boundaries and environment, 125–128; characteristics of political units, 128–133; conflict, causes of, 155–159; control of arms trade, 163–167; economics and the influence of arms trade, 127–128, 161–163; foreign policy and diplomacy, influenced by arms trade, 159–161; modes of interaction, 147–155; regime norms and rules, 155–167; structure and stratification, 133–147; summary of, 167–169
Postwar years (1992—): boundaries and environment, 196–197; characteristics of political units, 198; conflict, causes of, 202; control of arms trade, 203–205; economics and the influence of arms trade, 197, 202–203; foreign policy and diplomacy, influenced by arms trade, 202; modes of interaction, 201–202; regime norms and rules, 202–205; structure and stratification, 198–201
Producers of military equipment. *See* Suppliers
Production: co-, 94, 119, 203; internationalization of, 147–151, 201–202
Production, mode of: description of, 17, 52; Iraq example and, 20; 1930–40 period, 69–70, 71; 1946–66 period, 93–94; 1966–80 period, 109–110; 1980–92 period, 147–155; 1992- period, 201–202; subsystems, 146
Public Law 100–456 (U.S.), 177

-R-

Rand Corp., 145, 149
Reagan administration: dual-use equipment/technology and, 155; renewed arms production and, 125, 159, 163, 174, 188; research conducted on arms transfers and, 8
Recipient(s): behavior changes and responses from, 181–182; 1930–40 period, 64–69; 1946–66 period, 81–85, 86; 1980–92 period, 126, 133, 135; 1992- period, 201; as sources of data, 25; -supplier structures in 1930–40, 66, 68; -supplier structures in 1946–66 period, 86, 89; -supplier structures in 1966–80 period, 104–105, 106; -supplier structures in 1980–92 period, 138–139; systemic level and explaining behavior of, 170–172

Regime(s) norms and rules: defining, 53; East-West trade as a, 54; features of, 53–54; Holsti framework and, 53–56; 1930–40 period, 70, 72–75; 1946–66 period, 94–98; 1966–80 period, 110–121; 1980–92 period, 155–167; 1992– period, 202–205; nuclear proliferation as a, 54–55; U.S.-Soviet behavior as a, 54

Regional distribution: 1930–40 period, 68; 1946–66 period, 86, 90; 1966–80 period, 105, 107–108; 1980–92 period, 139; 1992– period, 197

Reliability, data, 34–36

Retransfer of weapons systems: 1930–40 period, 70; 1980–92 period, 147

-S-

Saudi Arabia, 159

Scud missile, 18–20, 158, 166

Second-tier suppliers, 130–132, 137–138, 179–181, 198

Secrecy, data reliability and, 35

Single European Act, 151, 202

South Africa and apartheid, influence of arms trade on, 2

South Korea, benefits of systemic knowledge and, 46

South Yemen, 157

Soviet Union: problems in measuring arms exports from, 22, 25, 34–35, 204; Afghanistan-Soviet conflict, 156–157, 190; future of, 199–200; as supplier, 84, 93, 94, 109, 113, 172–173, 183, 196–197; -United States behavior as a regime, 54; and structural influence 116–118, 157, 160

Sri Lanka, 157

*Statistical Year-book of the Trade in Arms and Ammunition,* 5–6, 23, 24

Stockholm International Peace Research Institute (SIPRI): *Arms Production in the Third World,* 8; *Arms Trade with the Third World, The,* 11, 12; *Arms Transfer Limitations and Third World Security,* 8–9; *Arms Transfers to the Third World: 1971–85,* 8–9, 206; data from, 104, 133, 134, 137, 139, 153–154; research and publications of, 8–9, 11–12, 24; *Yearbook on Armaments and Disarmament,* 26, 29–30

Strategic Defense Initiative, 125

Structure and stratification: Holsti framework and, 51; 1930–40 period, 64–69; 1946–66 period, 85–90; 1966–80 period, 103–108; 1980–92 period, 133–147; 1992– period, 198–201

*Structure of Scientific Revolutions, The,* (Kuhn), 12

*Sunday Times,* 184

Supplier(s)s: illegal versus private, 130; multinational corporations, 102, 132–33; 1930–40 period, 62–64; 1946–66 period, 79–85, 86; 1966–80 period, 100–103; 1980–92 period, 128–133, 135; 1992– period, 198–200; private versus public, 61; -recipient structures in 1930–40, 66, 68; -recipient structures in 1946–66 period, 86, 89; -recipient structures in 1966–80 period, 104–105, 106; -recipient structures in 1980–92 period, 138–139; second-tier, 130–132, 137–138, 179–181, 198; Soviet Union as, 172–173; subsidiaries created, 63–64; systemic level and explaining behavior of, 170–172; U.S. as, 173–176

Supplying governments as sources of data, 25–26

Surface-to-air missiles (SAMs), proliferation of, 143–144

Surface-to-surface missiles (SSMs), Scud, 183–184

Systemic approach: benefits of using, 170–171; utility of, 45–47

Systems and regimes: choosing among competing approaches and frameworks, 47–48; contradiction of 1980–92 period, 124–125; Holsti framework, 48–57; utility of systemic approach, 45–47. *See also* Regime(s) norms and rules

Systems transformations, continuity and change, 190–196

-T-

Technology: dual-use, 155; 1930–40 period, 61–62; 1946–66 period, 81; 1980–92 period, 128; future role, 201;
Terrorism, 113–14
Tornado aircraft, 21–22, 132, 133
Training and technical assistance: description of, 17; Iraq example and, 19
Transparency, 204
Transparency in Armaments resolution, 10, 200, 204
Turkey: -Cyprus conflict, 113; as a second-tier supplier, 179–180

-U-

United Kingdom, 2, 21–22, 82, 97, 139, 144, 147, 149, 156
United Nations Security Council, resolutions restricting arms exports and, 10, 205
United States: defense industries, 185–189; illegal arms exports from, 177–179; response to offsets, 176–77; -Soviet behavior as a regime, 54; as supplier, 173–176
U.S. Arms Control and Disarmament Agency (ACDA): data from, 85, 103, 133, 134, 137, 139; data reliability and, 34, 35–36; *World Military Expenditures and Arms Transfers* (WMEAT), 25, 27–29
U.S. Congress, Office of Technology Assessment, 128, 153, 174

U.S. Congress, role in arms exports, 2, 6, 7, 35, 38, 44, 174, 175, 177, 187, 188, 205
U.S. Defense Security Assistance Agency (DSAA), *Foreign Military Sales, Foreign Military Construction Sales and Military Assistance Facts*, 26, 30
U.S. Office of Management and Budget, 201
U.S. Senate Munitions Inquiry of 1934–36 (Nye committee hearings), 7, 72–73, 75

-V-

Validity: construct, 37; data, 36–40; face, 37
Vietnam War, 2

-W-

*Wall Street Journal*, 184
Warheads, 62
Warnke, Paul, 8
West European Union (WEU), 150
Wolfowitz, Paul, 189
World Bank, 128
*World Military Expenditures and Arms Transfers* (WMEAT)(ACDA), 25; data reliability and, 34, 35–36; dollar values recorded in, 27–29
World War I, disarmament and post-, 1
World War II, lend-lease agreements and, 2

-Y-

*Yearbook on Armaments and Disarmament* (SIPRI), 26; dollar values recorded in, 29–30
Yom Kippur War (1973), 2, 111–112, 113, 120

# About the Author

Edward J. Laurance graduated from the United States Military Academy in 1960. After ten years of military service, he began his academic career in 1970 with an M.A. from Temple University in political science, and a PhD. in 1973 from the University of Pennsylvania in international relations. In June of 1972, he accepted an appointment at the Naval Postgraduate School in Monterey, California, being promoted to associate professor with tenure in 1977, and full professor in 1990. In September 1991, he accepted a full-time appointment at the Monterey Institute of International Studies (MIIS), where he now teaches courses in international organizations and relations, military proliferation, and public policy analysis and research methods.

Professor Laurance has focused his research, teaching, and professional activities on the international arms trade. He has published articles in leading journals such as *The Journal of Conflict Resolution, International Studies Quarterly, Armed Forces and Society, Orbis,* and *Policy Sciences.* He has also written chapters for five books. He is a member of the International Studies Association, the Arms Control Association, the International Institute for Strategic Studies, and the Inter-University Seminar on Armed Forces and Society.

In 1978–79, Professor Laurance served in the U.S. Arms Control and Disarmament Agency as special assistant to the chief of the arms transfer branch. In 1980 and 1981, he testified on U.S. arms transfer policy and the U.S. security assistance program before the Subcommittee on International Security, Committee on Foreign Affairs, U.S. House of Representatives. He has lectured extensively on the subject of the international arms trade at U.S. war colleges, universities, and various public fora. In 1982, he was visiting professor in the department of political science at the University of California-Davis. From 1987 to 1989, he was the study director for a Rockefeller Foundation project to study the international arms trade. In September 1991, he was appointed co-director of the International Missile Proliferation Project at MIIS, a project which maintains and distributes a database on the motivations, capabilities, and patterns of trade of those states involved in the production, export, and acquisition of missiles in the developing world. In January 1992, he was appointed as a consultant to a United Nations panel charged with developing a database and other procedures in support of the U.N. resolution 46/36L of December 1991, which established an international arms trade register.